IEE ELECTROMAGNETIC WAVES SERIES 10

SERIES EDITORS: PROFESSOR P. J. B. CLARRICOATS,
G. MILLINGTON,
E. D. R. SHEARMAN
AND J. R. WAIT

APERTURE ANTENNAS AND DIFFRACTION THEORY

Previous volumes in this series

APERTURE ANTENNAS AND DIFFRACTION THEORY

E. V. Jull

Department of Electrical Engineering
The University of British Columbia
Vancouver
Canada

PETER PEREGRINUS LTD.
On behalf of the
Institution of Electrical Engineers

Published by: The Institution of Electrical Engineers, London
and New York
Peter Peregrinus Ltd., Stevenage, UK, and New York

British Library Cataloguing in Publication Data

Jull, Edward V.
    Aperture antennas and diffraction theory. —
    (Electromagnetic wave series; 10)
    1. Radio waves — Diffraction
    2. Antennas (Electronics)
    I. Title
    621.380'28'3   TK6553

ISBN 0-906048-52-4

Typeset at the Alden Press Oxford London and Northampton
Printed in England by A. Wheaton & Co., Ltd., Exeter

Contents

Preface

This book originated with an invitation to give a few lectures on aperture antennas to students during a stay at the Technical University of Denmark in 1964. The plane wave spectrum method of aperture analysis was adopted from the work of P.C. Clemmow and my former graduate supervisor J. Brown. It was then a new and refreshing change from the traditional approach. At about the same time, J.B. Keller visited the laboratory and introduced me to the geometrical theory of diffraction, from which has evolved an entirely new and in most respects more effective method of aperture antenna analysis. The ideas in this book owe much to these individuals.

The basis of the aperture analysis in Chapters 2–6 is, in essence, Kirchhoff's diffraction theory. In the remaining Chapters 7–9, the newer geometrical theory of diffraction is used. The analysis is almost entirely two-dimensional or scalar, with extensions to three dimensions indicated. In this way the essential concepts are conveyed in a simple form. Many practical antennas yield to a two-dimensional analysis as shown in the examples of Chapters 2 and 9. Also in at least the initial stages of antenna design it is usually convenient to assume a two-dimensional model.

As the book developed in response to teaching needs it contains some elementary material which may be helpful to those with no background in electromagnetic theory beyond Maxwell's equations. Much of Chapters 2–6 has benefitted from exposure to senior undergraduates at the University of British Columbia. Portions of Chapters 6–8 have been used as well in graduate courses at UBC, Carleton and Queen's universities.

Although intended primarily as a text, it is hoped the book may also be useful to practising engineers. Consequently, in the chapters dealing with applications, comparisons with experiment are made wherever possible. Examples of this are the horn pattern comparisons of Chapters 6 and 9. Most of these results were obtained by students and have not been published previously. The numerical data of Figs. 6.4 and 6.6 were obtained by Miss S.L. Paige and K.A. Grey and that of Fig. 6.5 by P.T.K. Chun. A. Safaai-Jazi obtained the geometrical diffraction theory patterns of Figs. 9.15 and 9.19 and N. Donatucci the experimental patterns.

Most of my contributions in this book were made in the Division of Electrical

Engineering of the National Research Council, Ottawa. The book was mainly written there also and I am grateful to its director J.Y. Wong for the generous provision of facilities and support during its preparation. Among many who helped at NRC was P. Amirault who expertly did most of the tracings. I am also indebted to B. Enander for the provision of facilities during a summer visit in the Division of Electromagnetic Theory of the Royal Institute of Technology, Stockholm. I have benefited as well from discussions with colleagues. In particular, S. Przeździecki was helpful with the uniform asymptotic theory of diffraction. Thanks are especially due to my friend and colleague R.A. Hurd who introduced me to rigorous diffraction theory years ago, arranged for our collaboration during a years' leave in which most of the writing was done and constructively criticised the typescript.

E.V. Jull
November 1980

Chapter 1

Introduction

An aperture antenna may be described in two ways. First, as an area of a surface with a radiating field distribution across it, the field being negligible on the surface outside the area. Second, as an area bounded by edges and excited by a source. Examples are radiating slots, horns and reflectors. These two descriptions characterise the two alternative methods of aperture antenna analysis used in this book. One is based on aperture field radiation, the other on aperture edge diffraction. Both apply strictly to large apertures so slots are generally excluded, but they are also useful for apertures with dimensions comparable to a wavelength. They tend to be complementary in that where one fails the other may succeed.

1.1 Historical background

The earliest practical antennas were wire structures operating at relatively low frequencies. Calculation of their radiated field is most conveniently based on the current on the structure. In principle the method is exact, but usually the current is not known exactly and approximations are made. With higher frequency communications, higher directivity antennas such as horns and reflectors were used. Analysis of these antennas in terms of the current on the structure is usually either very inconvenient or virtually impossible. Instead the radiated field in the forward direction is conventionally derived from the tangential field in the plane of the radiating aperture. This method is also exact in principle but essentially never in application for the field outside the aperture is usually assumed negligible and that in the aperture taken as the incident field from the feed. With these approximations the method is essentially that of optics called the Huygens-Kirchhoff method, or simply the Kirchhoff method, and is well established as accurate for the fields of apertures large in wavelengths.

The origins of this method can be traced to the ideas of Huygens who in 1690 proposed a geometrical theory of light propagation based on wavelets which expand radially at light velocity. Light intensity at any point is the envelope of contributions from the wavelets (see, for example, Baker and Copson, 1950). In 1818,

Fresnel included Huygens' ideas in a wave theory of diffraction. In it the diffracting object acts as a barrier to incident light and the resulting light intensity results from summing contributions of wavelets unobstructed by the barrier. Interference between secondary wavelets yields the diffraction pattern and the light intensity behind the barrier.

Another explanation of diffraction had been put forward by Young about 1804. He observed the position of diffraction fringes could be accounted for by interference between a direct light wave passing unobstructed through the aperture and a wave from the aperture edge. This agreed with the familiar observation that light in the shadow of a barrier appears to originate from the edge. Young's views on optics were evidently not well received generally at the time and with Fresnel's later demonstration that the shape of the diffracting edge had no apparent effect on the diffraction pattern, he abandoned his theory of diffraction in favour of Fresnel's.

The fate of Young's views seemed settled by the great success of solutions obtained from Kirchhoff's mathematical refinement of Fresnel's ideas in 1882 in accurately accounting for optical experiments. Kirchhoff's solution satisfied the scalar wave equation for light at all points in space, but it was difficult to see how Young's ideas ever could be properly accounted for mathematically across shadow boundaries where the incident wave must be discontinuous. An interesting account of all this has been given by Rubinowicz (1957).

The first exact diffraction solution, plane wave diffraction by a half-plane, was obtained by Sommerfeld (1896). Not surprisingly Sommerfeld's infinitely thin, perfectly conducting screen held little appeal to experimentalists in optics (e.g. Meyer, 1932), who have tended to adhere to the Kirchhoff theory, but it is a landmark in the newer field of radio science. At sufficient distance from the edges, Sommerfeld's solution for the total field divides into a geometrical optics wave and a diffracted wave which appears to originate from the edge. Schwarzschild (1902) obtained a solution for plane wave diffraction by an infinite slit on the basis of Sommerfeld's solution by considering successive interactions between the two half-planes which form the slit. For well separated half-planes, the diffraction pattern then becomes one of interference between the direct plane wave and diffracted cylindrical waves from the aperture edges.

The asymptotic forms of diffraction solutions which rigorously satisfied the scalar wave equation and boundary conditions thus supported the validity of Young's ideas on diffraction. Moreover, it was shown possible to split the field from Kirchhoff's surface integral over an aperture into an incident wave and a line integral along the aperture edge yielding diffracted waves in the far field (Rubinowicz, 1924), indicating an equivalence between Fresnel's and Young's explanation of diffraction.

Real growth in the use of Young's ideas has occurred in the last few decades as a result of developments in high-frequency diffraction theory. These developments have occurred mainly in the USA and the USSR. In the USA the outstanding proponent has been Keller (1953) who proposed a geometrical theory of diffraction which combines classical geometrical optics with asymptotic diffraction theory.

Diffraction is assumed to be an effect which occurs locally when an incident geometrical optics ray strikes an edge, producing diffracted rays in a manner similar to that in which reflected rays appear. The field on a diffracted ray is determined from the asymptotic form of the exact solution of the appropriate diffraction problem. Diffracted rays may in turn be diffracted and the total field is the sum of contributions from all rays, incident, reflected and diffracted. The mathematical basis of the method, its simplicity and clarity and Keller's work in extending and promoting the ideas have lead to a wide acceptance, Keller and Hansen (1965) have reviewed the developments which lead to the geometrical theory of diffraction.

Soviet developments in asymptotic diffraction theory are less centered on a single individual than in America. The Soviet literature, which is as impressive as it is extensive has, on the whole, made better use of the Western literature than the reverse. However, Ufimtsev's (1962) method of edge waves, or physical theory of diffraction, developed at about the same time as Keller's method and closely related to it, is also widely used to calculate high frequency diffracted fields. A helpful survey of Soviet contributions by Borovikov and Kinber (1974) includes antenna applications as well.

Applications of asymptotic diffraction theory to the analysis of aperture antennas began in the early 1960s and have increased since, until now the geometrical theory of diffraction or its equivalents has become the principal method for the analysis of antennas with dimensions large in wavelengths. Numerical methods based on the computer, which now dominate for small antennas, become unwieldy and expensive for large structures. While the extraordinary developments in computer technology are leading to their encroachment on the high-frequency regime, the superior conceptual and mathematical simplicity of the geometrical theory of diffraction assures its future in aperture antennas analysis and design.

1.2 Outline of the book

The method of aperture antenna analysis in Chapters 2–6 is based on Kirchhoff diffraction and Fourier transform theory. Its development in Chapter 2 from a plane wave spectrum field representation is a more recent departure from traditional methods. Earlier developments usually began with an integral representation between the fields at a point inside a region and those on a boundary surface containing the aperture. With fields assumed negligible everywhere else on the surface, integration reduced to the aperture itself. Stratton (1941, p. 464), for example, gives vector expressions for the fields in terms of those on the surface. Alternatively, scalar diffraction theory as described, for example, by Baker and Copson (1950) can be applied to each component of a vector field. Either way approximations such as are used in physical optics are then usually required to get the integrals into a form from which the field may be calculated. As Silver's (1949) derivation shows, in the lengthy process of deriving the diffraction integral the physical picture is obscured.

For planar apertures, the derivation in Chapter 2 is especially advantageous in illustrating the role of evanescent aperture fields and showing clearly the Fourier transform relationship between aperture fields. It uses the fact that any field can be represented by a superposition of plane waves with amplitudes which can be calculated from the tangential field, electric or magnetic, in the aperture plane. The plane wave spectrum representation of fields has been much used by Clemmow (1966) in diffraction solutions. Awareness of its advantages in aperture antenna analysis is increasingly evident in papers and in books such as Collin and Zucker (1969, Chap. 3) and Rhodes (1974).

Chapter 3 illustrates basic features of aperture antennas analysis and design using Fourier transforms. Pattern analysis of simple, compound and phase-shifted aperture distributions are given and a procedure indicated for pattern synthesis. An operational method of pattern analysis is mentioned.

Chapter 4 deals with near-field or Fresnel zone effects. With large apertures and short wavelengths these are often difficult to avoid. Here analytical solutions for the near fields of simple aperture distributions are given. Plane wave spectrum field representations conveniently include the effect of measuring antenna directivity. Far-field pattern prediction from near-field measurements, a subject of considerable current interest, is also discussed. Fourier transformation of patterns measured on a planar near-field surface is the simplest and most common method. Here, calculation of the far-field pattern from the modal coefficients of the near-field pattern measured on a cylindrical surface is described. As a result of improvements in computers, numerical techniques and instrumentation, spherical surface scanning now can be implemented also.

Chapter 5 begins with the familiar concepts of aperture gain and effective area and proceeds to the less comfortable ideas of supergain and aperture bandwidth. Again the plane wave spectrum is convenient in describing the radiative and reactive power of an aperture. The spectrum function, or radiation pattern, in principle specifies as well the antenna impedance. Rhodes (1974) demonstrates this for the planar dipole. His definition of aperture Q, the inverse of bandwidth, was shown to apply to this structure, but the practicality of its general application is questionable.

Chapter 5 concludes with derivations of near-field axial gain of circular and rectangular apertures with uniform and cosinusoidal distributions. As the rectangular aperture expressions are essentially the same as those encountered later for pyramidal horns it is convenient to introduce and tabulate the functions encountered.

Applications of Kirchhoff diffraction theory to a few practical antennas are given in Chapter 6. The simple example of an open-ended waveguide set in a conducting plane shows the method is applicable even to apertures small in wavelengths provided the aperture plane boundary conditions are well satisfied. Pyramidal and sectoral horns are considered in some detail. Pyramidal horns are perhaps the earliest type of microwave antenna and are still widely used, yet evidence of the accuracy of their pattern prediction by the Kirchhoff method has not been generally available. The results given here show good accuracy for standard horns, even for angles far off the beam axis. Near-field patterns of these horns can be approximated by a minor change of parameters in the far field pattern expressions. Schelkunoff's reliable result for the axial gain follows simply from expressions for the axial near field gain of a rectangular aperture. When E-plane pattern and gain expressions are examined the effect of the narrow aperture dimension on their accuracy is evident. The conventional gain equation is seriously in error and a better one suggested.

Paraboloidal reflectors are introduced in Chapter 6, but considered only so far as to indicate how the previous methods may be applied. This assists the subsequent analysis of the horn reflector, an antenna also widely used in microwave communications.

The reader may conclude from Chapter 6 that the Kirchhoff method is usually adequate for field calculations in front of apertures. For fields in the lateral and rear directions, it is convenient to assume diffraction is an edge, rather than an aperture, effect. This the geometrical theory of diffraction does and, as a basis for it, Chapter 7 outlines rigorous diffraction theory for conductors with sharp edges. A full description of the main canonical solution, plane wave diffraction by a half-plane, is given with details in the appendices. Generalisations to oblique incidence, local sources and the wedge are then presented. Thus the tools for the final chapters are assembled.

Although the geometrical theory of diffraction is widely used, descriptions in books are few. Welcome recent contributions are chapters by Kouyoumjian (1975) and Jones (1979) and books by James (1976) and, in Russian, by Borovikov and Kinber (1978). The account in Chapter 8 is restricted to conductors with sharp edges in accordance with the subject of this book. Following Keller's (1957) example, diffraction by a slit illustrates the method and the circular aperture provides an example for curved edges. In these solutions shadow boundary singularities, a major difficulty with Keller's method, fortuitously self-cancel. Procedures for overcoming this difficulty in more usual circumstances are mentioned at the end of the chapter. These include the uniform geometrical theory of diffraction and the uniform asymptotic theory of diffraction, each with its merits and limitations.

Applications of geometrical diffraction theory are given in Chapter 9. To begin, the open-ended parallel-plate waveguide "antenna" is chosen for a convenient test of the method against an exact solution. Yee and Felsen's (1968) ray-optical analysis of reflection and radiation from this aperture provides a measure of the significance of higher-order edge interaction. The ray-optical method is essentially Keller's method with shadow boundary singularities approximately accounted for. A comparison with the precise uniform asymptotic theory solution is made. Numerical comparisons with exact results have demonstrated that the ray-optical method, which strictly applies to large apertures, is capable of accuracy for aperture widths as small as a few tenths of a wavelength for TEM mode incidence and for aperture widths just above TE_1 mode cutoff.

In the remainder of Chapter 9 the expression for the far field of a line source parallel and near to the edge of a conducting half-plane is used to avoid the shadow and reflection boundary singularities inherent in a direct application of the geometrical theory of diffraction. The expression is exact in the far field and simpler to use than the uniform asymptotic theory. The first example is the simplest of reflector antennas — a dipole in front of a conducting strip. Geometrical diffraction theory is shown to be particularly effective in providing its entire H-plane pattern. The H-plane fields of finite corner reflector antennas are similarly derived following Ohba's (1963) analysis. With a parabolic cylinder reflector only lateral and rear H-plane fields are derived as the forward pattern is more easily and usually adequately provided by the Kirchhoff method.

E-plane pattern analysis of pyramidal horns has also proven to be particularly well suited to a geometrical diffraction approach. Adopting the model of Russo *et al.* (1965) the waveguide feed is replaced by a magnetic line source and the analysis follows that of Yu *et al.* (1966). Comparisons between numerical and experimental results show interactions between the horn edges and its interior surfaces usually may be omitted in calculating the pattern in front of the horn and then it suffices to use the Kirchhoff result, which is almost the same. Geometrical diffraction theory is needed only to provide the lateral and rear field. There, first-order edge interaction is necessary for more than just the general level of backradiation.

As pyramidal horns are microwave gain standards a high accuracy in their predicted gain is desirable. The major source of error in this gain prediction is interaction between the aperture edges and the interior horn surfaces. Most of it occurs in the E-plane and can be estimated by geometrical diffraction theory. This is done near the end of Chapter 9. Finally, the complex reflection coefficient of an E-plane sectoral horn is calculated. As expected, the results are adequate if horn apertures are sufficiently large and waveguide feeds sufficiently small.

The appendixes contain a brief exposition of some mathematical techniques used together with details of the half-plane diffraction solution and a derivation of the transmission cross-section of an aperture.

Chapter 2

Plane waves from apertures

2.1 Plane wave solutions of Maxwell's equations

In free space, Maxwell's equations, which relate spatial and time variations of the electric and magnetic field intensities $\bar{E}$ and $\bar{H}$, are

$$\nabla \times \bar{E} = -\mu_0(\partial \bar{H}/\partial t) \tag{2.1}$$

$$\nabla \times \bar{H} = \epsilon_0(\partial \bar{E}/\partial t) \tag{2.2}$$

and

$$\nabla \cdot \bar{E} = \nabla \cdot \bar{H} = 0 \tag{2.3}$$

where ϵ_0 and μ_0 are the free space permittivity and permeability. Combining and eliminating $\bar{H}$ yields

$$\nabla^2 \bar{E} - \mu_0 \epsilon_0 (\partial^2 \bar{E}/\partial t^2) = 0 \tag{2.4}$$

and similarly for $\bar{H}$. From any field solution for $\bar{E}$ or $\bar{H}$, another can be obtained by substituting $\bar{E}$ for $\bar{H}$, $\bar{H}$ for $-\bar{E}$ and ϵ_0 for μ_0.

In Cartesian coordinates, eqn. 2.4 is

$$\nabla^2 E_x - \mu_0 \epsilon_0 (\partial^2 E_x/\partial t^2) = 0 \tag{2.5}$$

and similarly for E_y and E_z. If E_x varies with z only, we have

$$\frac{\partial^2 E_x}{\partial z^2} = \mu_0 \epsilon_0 \frac{\partial^2 E_x}{\partial t^2}. \tag{2.6}$$

A harmonic time variation of the fields is usually assumed.† If all fields vary with time t as $\exp(j\omega t)$, where $\omega = 2\pi f$ is the angular frequency, then $\partial/\partial t$ may be

---

†An arbitrary time variation of the field $E_x(t)$ can be expressed as the sum of a Fourier series of harmonic time variations, each harmonic having an amplitude $a(\omega)$. Thus

$$E_x(t) = \int_{-\infty}^{\infty} a(\omega) \exp(j\omega t)\, d\omega$$

where

$$a(\omega) = 1/(2\pi) \int_{-\infty}^{\infty} E_x(t) \exp(-j\omega t)\, dt$$

replaced by $j\omega$. A solution of eqn. 2.6 is then

$$E_x = A \exp[-jk(z - ct)] + B \exp[jk(z + ct)] \qquad (2.7)$$

where $k = \omega \sqrt{\mu_0 \epsilon_0}$ is the free space propagation constant, $c = \omega/k = (\mu_0 \epsilon_0)^{-1/2}$ is the free space wave velocity and A and B are real constants. The first term in eqn. 2.7 remains constant if z increases as ct increases and so represents a wave in the $+z$ direction. The second term represents a wave in the $-z$ direction. This complex exponential form of spatial and time variation of fields is used for convenience. It is understood that actual field quantities are the real part; i.e. for the plane wave in the z-direction the field is

$$E_x = \mathrm{Re}\{A \exp[-j(kz - \omega t)]\} = A \cos(kz - \omega t) \qquad (2.8)$$

In complex exponential notation, the time factor $\exp(j\omega t)$ is usually suppressed. Time-averaged power flux density is conveniently obtained from $\frac{1}{2}\mathrm{Re}(\bar{E} \times \bar{H}^*)$, where $*$ indicates complex conjugate.

2.2 General plane wave solutions

For a plane wave in the direction ζ, fields behave as $\exp(-jk\zeta)$. With ζ measured

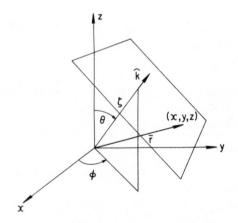

Fig. 2.1   *Coordinates for plane wave propagation*

from the origin of Fig. 2.1, and with direction cosines $\sin\theta \cos\phi$, $\sin\theta \sin\phi$ and $\cos\theta$,

$$\zeta = x \sin\theta \cos\phi + y \sin\theta \sin\phi + z \cos\theta \qquad (2.9)$$

which defines a plane surface. A unit vector normal to this plane is

$$\hat{k} = \bar{k}/k = \hat{x} \sin\theta \cos\phi + \hat{y} \sin\theta \sin\phi + \hat{z} \cos\theta \qquad (2.10)$$

where $\bar{k}$ is the vector propagation constant. Since $\bar{r} = \hat{x}x + \hat{y}y + \hat{z}z$, $\exp(-jk\zeta)$ can be written

$$\exp(-j\bar{k} \cdot \bar{r}) = \exp[-jk(x \sin\theta \cos\phi + y \sin\theta \sin\phi + z \cos\theta)] \quad (2.11)$$

For plane waves of this form, with spatial dependence of the fields on ζ only, the operator $\nabla = \hat{k}\partial/\partial\zeta$ can be replaced by $-j\bar{k}$ in Maxwell's equations. Then eqns. 2.1–2.3 become

$$\bar{k} \times \bar{E} = \omega\mu_0\bar{H} \quad (2.12)$$

$$-\bar{k} \times \bar{H} = \omega\epsilon_0\bar{E} \quad (2.13)$$

$$\bar{k} \cdot \bar{E} = \bar{k} \cdot \bar{H} = 0 \quad (2.14)$$

Hence the vectors $\bar{E}$, $\bar{H}$ and $\bar{k}$ form a right-handed set and for plane waves $\bar{E}$ and $\bar{H}$ are always orthogonal to $\bar{k}$. Also $\bar{k} \times$ eqn. 2.12 combined with eqns. 2.13 and 2.14 yields $k = \omega \sqrt{\mu_0\epsilon_0}$, the propagation constant magnitude, which when used in eqn. 2.12 gives

$$\hat{k} \times \bar{E} = Z_0\bar{H}, \quad (2.15)$$

where $Z_0 = \sqrt{\mu_0/\epsilon_0} \cong 120\pi \, \Omega$ is the wave impedance of free space. From eqn. 2.15, time-averaged power flux in the direction of propagation is $\frac{1}{2}\mathrm{Re}(\bar{E} \times \bar{H}^*) \cdot \hat{k} = \frac{1}{2}Y_0|E|^2$, where $Y_0 = Z_0^{-1}$ is the free-space plane wave admittance.

2.3 Homogeneous and inhomogeneous plane waves

Plane waves are 'homogeneous' if their equiamplitude surfaces, which are parallel planes, are also equiphase surfaces. To illustrate, consider a two-dimensional field uniform in, say, the y-direction. A plane wave of unit amplitude in the direction α of Fig. 2.2 may be written

$$\exp[-jk(x \sin\alpha + z \cos\alpha)] = \exp[-jkr \cos(\theta - \alpha)] \quad (2.16)$$

where $x = r \sin\theta, z = r \cos\theta$. For real angles α this wave is homogeneous. In what follows we shall want to include direction angles which are complex. Suppose $\alpha = \alpha_1 + j\alpha_2$ where α_1 and α_2 are real. Then eqn. 2.16 becomes the inhomogeneous plane wave

$$\exp[-jkr \cosh\alpha_2 \cos(\theta - \alpha_1)] \exp[kr \sinh\alpha_2 \sin(\theta - \alpha_1)] \quad (2.17)$$

which propagates in a direction α_1 from the z-axis and decays exponentially at right angles to this direction. To see this, let $\alpha_1 = \pi/2$. Then the wave is

$$\exp(-jkx \cosh\alpha_2) \exp(-kz \sinh\alpha_2) \quad (2.18)$$

which clearly behaves in the manner described.

A wave in which equiamplitude and equiphase surfaces do not coincide is called inhomogeneous. Such waves are often able to propagate along a plane boundary between two media; then they are usually called surface waves. Another situation in which inhomogeneous plane waves are found is a waveguide operating at a frequency below the cut-off frequency. Waves then propagate across the guide at the input and are exponentially attenuated down the guide axis.

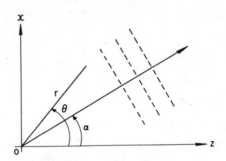

Fig. 2.2   Homogeneous plane wave
The direction of propagation is indicated by the arrow and equiamplitude and equiphase surfaces by the dashed lines. The equiamplitude surfaces of an inhomogeneous plane wave are not orthogonal to the direction of propagation.

2.4. Plane wave spectrum

The approach described in the balance of this chapter for the analysis of aperture antennas appears to have originated with Woodward and Lawson (1948) and Booker and Clemmow (1950) for two-dimensional fields and was extended to three dimensions by Brown (1958), Collin and Rothschild (1963) and Rhodes (1964). The plane wave spectrum representation of electromagnetic fields is well described by Clemmow (1966).

All fields are assumed to be uniform in the y-direction of Fig. 2.2 and it is assumed that the electric field lies entirely in the $x-z$ plane. Then any plane wave radiated in the direction α from an aperture in the $z = 0$ plane has an x-component of electric field given by

$$E_x(x,z) = A(\alpha) \exp[-jk(x \sin \alpha + z \cos \alpha)] \tag{2.19}$$

where $A(\alpha)$ is the amplitude of the wave. Imagine a collection of such plane waves propagating in all directions α from the $z = 0$ plane with an amplitude $A(\alpha)$ for each direction of propagation. The field equations are linear so each wave may be added and the total field got by summing over all directions α. It is convenient to let $k_x = k \sin \alpha$, $k_z = k \cos \alpha$ and $F(k_x) = \lambda A(\alpha)$ in eqn 2.19. The total E_x field in $z \geqslant 0$ is then

$$E_x(x,z) = \frac{1}{2\pi} \int_{-\infty}^{\infty} F(k_x) \exp[-j(k_x x + k_z z)] \, dk_x \tag{2.20}$$

where

$$k_z = \sqrt{k^2 - k_x^2} \text{ when } k_x^2 \leqslant k^2 \tag{2.21}$$
$$= -j\sqrt{k_x^2 - k^2} \text{ when } k_x^2 > k^2$$

All real values of k are contained in eqn. 2.20 so that homogeneous and inhomogeneous plane waves are included. For the former $k_x^2 < k^2$ and the positive root is chosen for k_z in order that the waves propagate outward in the z-direction. The latter, for which $k_x^2 > k^2$, are waves which propagate at complex angles. With k_z negative imaginary, they decay exponentially with increasing z. These evanescent waves traverse the aperture plane but do not carry energy away. They affect the reactance, rather than the resistance, of the antenna.

Putting $z = 0$ in eqn. 2.20 gives the aperture field required to sustain the x-component of the total field

$$E_x(x,0) = \frac{1}{2\pi} \int_{-\infty}^{\infty} F(k_x) \exp(-jk_x x) \, dk_x \tag{2.22}$$

$F(k_x)$ is usually called the angular spectrum of the field. From the form of eqn. 2.22 it is the inverse Fourier transform (see Appendix A.1) of the aperture distribution and may be written

$$F(k_x) = \int_{-\infty}^{\infty} E_x(x,0) \exp(jk_x x) \, dx \tag{2.23}$$

The electric field vector of a plane wave is normal to the direction of propagation $\hat{k}$. Thus each plane wave here will generally have both x and z components, whose relative magnitudes may be calculated from $\hat{k} \cdot \overline{E} = 0$. The total electric field is

$$\overline{E}(x,z) = \frac{1}{2\pi} \int_{-\infty}^{\infty} \left(\hat{x} - \frac{k_x}{k_z} \hat{z} \right) F(k_x) \exp[-j(k_x x + k_z z)] \, dk_x \tag{2.24}$$

and since the magnetic field of each plane wave is $Y_0 \hat{k} \times \overline{E}$, from eqn. 2.15,

$$\overline{H}(x,z) = \frac{\hat{y} Y_0 k}{2\pi} \int_{-\infty}^{\infty} \frac{F(k_x)}{k_z} \exp[-j(k_x x + k_z z)] \, dk_x \tag{2.25}$$

2.5 Radiation pattern

If the radiated field of eqn. 2.24 is observed at a large distance from the aperture, and the aperture field is confined to a relatively small length of the x-axis, waves reaching a distance point r,θ travel in essentially parallel lines at an angle θ to the z-axis (see Fig. 2.3). Only waves with phase differences which are multiples of 2π radians will interfere constructively; i.e. for constructive interference, $kd \sin \theta = 2\pi n$, $n = 1, 3, \ldots$. The smallest value of d is $\lambda/\sin \theta$ and hence λ is the smallest value of the period d for real angles θ. It follows that aperture field periodicities smaller than λ alter not the radiation pattern, but only the evanescent fields in the aperture.

In polar co-ordinates r,θ, eqn. 2.24 becomes

$$\bar{E}(r,\theta) = \frac{1}{\lambda} \int_C [\hat{r} \sin(\theta - \alpha) + \hat{\theta} \cos(\theta - \alpha)] F(k \sin \alpha)$$

$$\exp[-jkr \cos(\theta - \alpha)] \, d\alpha \tag{2.26}$$

where the integration contour C ranges over values of α from $-\pi/2 - j\infty$ to $\pi/2 + j\infty$ as shown in Fig. 2.4. The field of eqn. 2.26 at a distant point can be calculated by the stationary phase method of integration (see Appendix A.2). Here the physical

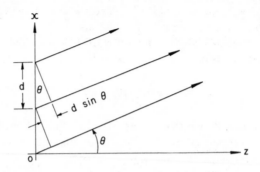

Fig. 2.3   *Propagation directions for calculation of the far-field radiation pattern of an aperture in the $z = 0$ plane*

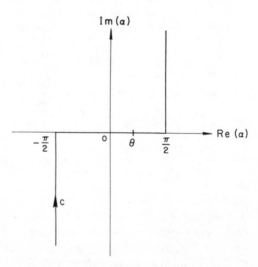

Fig. 2.4   *Integration contour in the complex α-plane*

arguments on which this method is based are sufficient. If $kr \gg 1$, the amplitudes of the inhomogeneous waves which arise from that part of C on which α is imaginary are, from eqn. 2.17, negligibly small. Also the exponential part of the integrand oscillates rapidly for all values of α except those near $\alpha = \theta$. Physically this means there is destructive interference of all waves except those in the direction $\alpha = \theta$.

Thus the nonexponential part of the integrand may be replaced by its value for $\alpha = \theta$, yielding

$$\overline{E}(r,\theta) = \frac{\hat{\theta}}{\lambda} F(k \sin \theta) \int_C \exp[-jkr \cos(\theta - \alpha)] \, d\alpha$$

$$= \hat{\theta} \frac{\pi}{\lambda} F(k \sin \theta) H_0^{(2)}(kr) \tag{2.27}$$

where the integral representation of the zero-order Hankel function of the second kind $H_0^{(2)}$ (kr) (see Appendix A.3) has been recognised. Its asymptotic form for $kr \gg 1$,

$$H_0^{(2)}(kr) \simeq \sqrt{2/(\pi kr)} \, \exp[-j(kr - \pi/4)] \tag{2.28}$$

in eqn. 2.27 gives

$$\overline{E}(r,\theta) \simeq \hat{\theta} \frac{F(k \sin \theta)}{\sqrt{r\lambda}} \exp[-j(kr - \pi/4)]$$

$$= \frac{\hat{\theta} \exp[-j(kr - \pi/4)]}{\sqrt{r\lambda}} \int_{-\infty}^{\infty} E_x(x,0) \exp(jkx \sin \theta) dx \tag{2.29}$$

Eqns. 2.27 and 2.29 represent a cylindrical wave and $F(k \sin \theta)$ gives its variation with θ. Thus it is evident that in the far field the angular spectrum is the radiation pattern of the aperture. This is an extremely useful result for it is now radiation pattern and aperture distribution which are a Fourier transform pair and Fourier transforms may be applied directly to the calculation of antenna radiation patterns.

If the aperture electric field was polarised in the y-direction of Fig. 2.2, we would define $F_y(k_x)$ as the Fourier transform of $E_y(x,0)$ and obtain the far radiation field

$$\overline{E}(r,\theta) \simeq \hat{y} \frac{\cos \theta \, \exp[-j(kr - \pi/4)]}{\sqrt{r\lambda}} \int_{-\infty}^{\infty} E_y(x,0) \exp(jkx \sin \theta) dx \tag{2.30}$$

An arbitrarily polarised two-dimensional aperture field can be resolved into transverse electric (TE) or E-polarised and transverse magnetic (TM) or H-polarised components. Eqn. 2.30 represents the far field of the TE component radiating into $z > 0$ and eqn. 2.29 the far field of the TM component. The total far field will be their vector sum.

Although integration limits above are infinite, in practice they correspond to the aperture edges, it being assumed fields vanish in $z = 0$ outside a finite length of the x-axis. The radiation pattern is the Fourier transform of the aperture field provided the distance r is large in terms of the aperture dimensions and the wavelength. Of course the same convenient relationship holds for a three-dimensional system of plane waves.

2.6 Plane wave spectra in three dimensions

Again two spectrum functions are required to specify an arbitrarily polarised field radiated from an aperture. These are Fourier transforms of the two orthogonal components of aperture field, electric or magnetic. We shall deal with fields polarised in the $x-z$ plane of Fig. 2.5 first. These have no y-component of electric field.

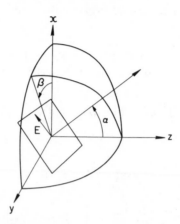

Fig. 2.5   *Plane wave radiating from an aperture in the $z = 0$ plane*

The x-component of a plane wave in the direction α, β can be written

$$E_x(x,y,z) = \lambda^2 / F(k \sin \alpha \cos \beta, \ k \sin \alpha \sin \beta)$$

$$\exp[-jk(x \sin \alpha \cos \beta + y \sin \alpha \sin \beta + z \cos \alpha)] \qquad (2.31)$$

Let $k_x = k \sin \alpha \cos \beta$, $k_y = k \sin \alpha \sin \beta$ and $k_z = k \cos \alpha$. Then the x-component of the total field for the complete spectrum of plane waves is

$$E_x(x,y,z) = \frac{1}{4\pi^2} \int_{-\infty}^{\infty} \int_{-\infty}^{\infty} F(k_x,k_y) \exp[-j(k_x x + k_y y + k_z z)] \, dk_x dk_y \qquad (2.32)$$

where

$$k_z = \begin{cases} \sqrt{k^2 - (k_x^2 + k_y^2)}, & \text{when } k_x^2 + k_y^2 \leqslant k^2 \\ -j \sqrt{(k_x^2 + k_y^2) - k^2}, & \text{when } k_x^2 + k_y^2 > k^2 \end{cases} \qquad (2.33)$$

Waves for which $k_x^2 + k_y^2 \leqslant k^2$, corresponding to real angles α, β, contribute to the radiation field of the aperture, whereas those for which $k_x^2 + k_y^2 > k^2$ correspond to complex angles α, β and form the evanescent field of the antenna aperture.

From eqn. 2.32 the aperture field is

$$E_x(x,y,0) = \frac{1}{4\pi^2} \int_{-\infty}^{\infty} \int_{-\infty}^{\infty} F(k_x,k_y) \exp[-j(k_x x + k_y y)] \, dk_x dk_y \qquad (2.34)$$

where

$$F(k_x,k_y) = \int_{-\infty}^{\infty} \int_{-\infty}^{\infty} E_x(x,y,0) \exp[j(k_x x + k_y y)] \, dx\, dy \qquad (2.35)$$

The z-component of $\bar{E}$ in each plane wave follows from eqn. 2.14; i.e.

$$\bar{k} \cdot \bar{E} = (\hat{x} k_x + \hat{y} k_y + \hat{z} k_z) \cdot (\hat{x} E_x + \hat{z} E_z) = 0$$

or

$$E_z(x,y,0) = -\frac{k_x}{k_z} E_x(x,y,0)$$

and the complete $\bar{E}$ field is

$$\bar{E}(x,y,z) = \frac{1}{4\pi^2} \int_{-\infty}^{\infty} \int_{-\infty}^{\infty} (\hat{x} k_z - \hat{z} k_x) F(k_x,k_y)$$

$$\exp[-j(k_x x + k_y y + k_z z)] \, \frac{dk_x dk_y}{k_z} \qquad (2.36)$$

The magnetic field components of each plane wave in the aperture follow from $\bar{H} = Y_0 \hat{k} \times \bar{E}$ with $E_y = 0$. Then the magnetic field is

$$\bar{H}(x,y,z) = \frac{Y_0}{4\pi^2 k} \int_{-\infty}^{\infty} \int_{-\infty}^{\infty} (-\hat{x} k_y k_x + \hat{y}\, (k^2 - k_y^2) - \hat{z} k_y k_z)$$

$$\times F(k_x,k_y) \exp[-j(k_x x + k_y y + k_z z)] \, \frac{dk_x dk_y}{k_z} \qquad (2.37)$$

To evaluate eqn. 2.36 for large kr, we first convert to spherical co-ordinates, then apply stationary phase integration to the double integral, which has a stationary point at $\alpha = \theta$, $\beta = \phi$, as in Appendix A.2.2. The final result is

$$\bar{E}(r,\theta,\phi) \simeq \frac{j \exp(-jkr)}{\lambda r} (\hat{\theta} \cos \phi - \hat{\phi} \sin \phi \cos \theta) F(k \sin \theta \sin \phi) \qquad (2.38)$$

and

$$\bar{H}(r,\theta,\phi) = Y_0 \hat{r} \times \bar{E}(r,\theta,\phi), \qquad (2.39)$$

In eqn. 2.38 $F(k_x,k_y)$ is defined by eqn. 2.35 as the Fourier transform of the tangential electric field in the aperture. Alternatively, we could begin with a magnetic field in the aperture plane with no x-component and define (Smith, 1963)

$$F(k_x,k_y) = Z_0 \int_{-\infty}^{\infty} \int_{-\infty}^{\infty} H_y(x,y,0) \exp[j(k_x x + k_y y)] \, dx\, dy \qquad (2.40)$$

We get, for large kr,

$$\bar{E}(r,\theta,\phi) = \frac{j \exp(-jkr)}{\lambda r} (\hat{\theta} \cos \phi \cos \theta - \hat{\phi} \sin \phi) F(k \sin \theta \cos \phi,$$

$$k \sin \theta \sin \phi) \qquad (2.41)$$

and $\bar{H}(r,\theta,\phi)$ given by eqn. 2.39. This is an alternative representation of the radiation field to eqns. 2.38 and 2.39. For small angles θ, $\cos \theta \simeq 1$ and the two representations are identical. The Huygens-Kirchhoff method calculating the diffraction pattern of an aperture as given, for example, by Silver (1949, p. 165) requires that both electric and magnetic fields in the aperture plane are specified. Assuming $E_x = Z_0 H_y$ in the aperture, we can get this result by superimposing eqns. 2.38 and 2.41, giving

$$\bar{E}(r,\theta,\phi) = \frac{j \exp(-jkr)}{2\lambda r}(1 + \cos \theta)(\hat{\theta} \cos \phi - \hat{\phi} \sin \phi)$$

$$\times F(k \sin \theta \cos \phi, k \sin \theta \sin \phi) \qquad (2.42)$$

where F is given by eqn 2.35.

The results of eqns. 2.38, 2.40 and 2.42 may be compared with corresponding results for electromagnetic wave diffraction by an aperture in a plane conducting screen with a unit normal $\hat{n}$. Eqn. 2.38 could have been obtained by calculating the field of a fictitious magnetic current sheet of density $\bar{M} = \bar{E} \times \hat{n}$ in the aperture with $\bar{E}$ the incident electric field. Putting $\bar{M} = 0$ in the $z = 0$ plane outside the aperture satisfies the boundary condition on the conductor, but there is an approximation in assuming that the aperture tangential electric field is its undisturbed value. Eqn. 2.41 is obtained when an electric current sheet $\bar{J} = \hat{n} \times \bar{H}$, is assumed in the aperture, with $\bar{H}$ the incident magnetic field. This is in accordance with the tangential $\bar{H}$ being undisturbed in the aperture but $\hat{n} \times \bar{H} = 0$ on the conductor is not a valid boundary condition. Eqn. 2.42 follows if both electric and magnetic current sheets are present in the aperture with aperture $\bar{E}$ and $\bar{H}$ related by normal plane wave conditions and $\bar{E} \times \hat{n} = \hat{n} \times \bar{H} = 0$ for $z = 0$ outside the aperture. In this formulation both $\bar{E} \times \hat{n} = 0$ in the aperture and $\hat{n} \times \bar{H} = 0$ on the conductor are incorrect boundary conditions, but Huygens' principle, which requires that there be no backward radiation of the wavefront, is satisfied. Jordan and Balmain (1968, p. 488) show it is equivalent to assuming that a TEM field exists in the aperture plane and radiates into $z > 0$.

For aperture antennas such as slots in a conducting screen the radiation pattern can be obtained from the aperture electric field alone and eqn. 2.38 is the correct expression to use. The aperture field of an antenna is rarely known with much precision, however, and for apertures not in a conducting screen there is usually little to chose between eqns. 2.38, 2.41 and 2.42. All three provide accurate results for small angles θ, where they are essentially identical, and have less precision at angles far off the beam axis. Eqn. 2.42, which combines both electric and magnetic wave types, is most commonly adopted.

2.7 Arbitrarily polarised aperture fields

In Section 2.6, the aperture field was assumed to be linearly polarised in the x–z plane, with no E_y component. For aperture fields polarised in the y–z plane, i.e. with no E_x component, the radiated fields can be written

$$\overline{E}(x,y,z) = \frac{1}{4\pi^2} \int_{-\infty}^{\infty} \int_{-\infty}^{\infty} (\hat{y}k_z - \hat{z}k_y)F_y(k_x,k_y)$$

$$\exp[-j(k_x x + k_y y + k_z z)] \frac{dk_x dk_y}{k_z} \tag{2.43}$$

where $F_y(k_x,k_y)$ is the double Fourier transform of $E_y(x,y,0)$. For $kr \gg 1$, eqn. 2.43 becomes

$$\overline{E}(r,\theta,\phi) \simeq \frac{j \exp(-jkr)}{\lambda r}(\hat{\theta} \sin\phi + \hat{\phi} \cos\theta \cos\phi)F_y(k \sin\theta \cos\phi,$$

$$k \sin\theta \sin\phi) \tag{2.44}$$

For an arbitrarily polarised aperture field, the radiation field will be the sum of eqns. 2.36 and 2.43. In the far field this is

$$\overline{E}(r,\theta,\phi) \simeq \frac{j \exp(-jkr)}{\lambda r}\{(\hat{\theta} \cos\phi - \hat{\phi} \sin\phi \cos\theta)F_x(k \sin\theta \cos\phi,$$

$$k \sin\theta \sin\phi) + (\hat{\theta} \sin\phi + \hat{\phi} \cos\phi \cos\theta)F_y(k \sin\theta \cos\phi,$$

$$k \sin\theta \sin\phi)\} \tag{2.45}$$

where

$$F_x = \int_{-\infty}^{\infty} \int_{-\infty}^{\infty} E_x(x,y,0) \exp[jk(x \sin\theta \cos\phi + y \sin\theta \sin\phi)] \, dxdy \tag{2.46}$$

$$F_y = \int_{-\infty}^{\infty} \int_{-\infty}^{\infty} E_y(x,y,0) \exp[jk(x \sin\theta \cos\phi + y \sin\theta \sin\phi)] \, dxdy \tag{2.47}$$

If we derive our radiation field from both tangential electric and tangential magnetic fields in the aperture, we obtain, in place of eqn. 2.45,

$$\overline{E}(r,\theta,\phi) \simeq \frac{j \exp(-jkr)}{\lambda r}(1 + \cos\theta)\{(\hat{\theta} \cos\phi - \hat{\phi} \sin\phi)F_x$$

$$+ (\hat{\theta} \sin\phi + \hat{\phi} \cos\phi)F_y\} \tag{2.48}$$

These expressions are in accordance with the general result that the fields of any source can be expressed in terms of two scalar functions, in this case the Fourier transforms of two orthogonal components of aperture field. Further developments in this approach to the analysis and synthesis of aperture antennas have been given by Rhodes (1974).

Chapter 3

Fourier transform representation of aperture patterns

3.1 Separable aperture distribution

The Fourier transform relationship between an aperture distribution and its radiation pattern is exact, but its application involves an approximation at once for the field is never known over the entire aperture plane. Usually it is known approximately only within the aperture and is assumed to vanish in the rest of the aperture plane. Suppose a field exists in an aperture S in the plane $z = 0$ and is negligible outside S. The tangential electric field in S is resolved into appropriate components, usually Cartesian if the aperture shape is rectangular, as in Fig. 3.1. Then eqns. 2.38 or 2.42 gives the radiation field with integration over S only. Here we consider

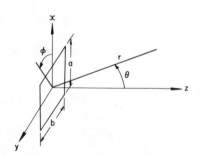

Fig. 3.1   *Coordinates for radiation from a rectangular aperture*

only the aperture electric field $E(x,y)$ polarised in the x-direction; the other component, if present, can be dealt with separately. From eqn. 2.42 the electric field at r,θ,ϕ is

$$\bar{E}(r,\theta,\phi) = \bar{A} \int_S \int E(x,y) \exp[jk(x \sin\theta \cos\phi + y \sin\theta \sin\phi)] \, dx \, dy$$

where

$$(3.1)$$

$$\bar{A} = \bar{A}(r,\theta,\phi) = \frac{j \exp(-jkr)}{2\lambda r}(1 + \cos\theta)(\hat{\theta}\cos\phi - \hat{\phi}\sin\phi) \qquad (3.2)$$

In most antenna designs, $E(x,y)$, in general complex, is separable in amplitude and phase in terms of the aperture coordinates. Otherwise little can be done analytically with eqn. 3.1. With $E(x,y) = E_1(x)E_2(y)$, eqn. 3.1 can be written

$$\bar{E}(r,\theta,\phi) = \bar{A} F_1(k_1) F_2(k_2) \tag{3.3}$$

where

$$F_1(k_1) = \int E_1(x) \exp(jk_1 x) dx$$
$$F_2(k_2) = \int E_2(y) \exp(jk_2 y) dy \tag{3.4}$$

and

$$\left.\begin{array}{l} k_1 = k \sin\theta \cos\phi \\ k_2 = k \sin\theta \sin\phi \end{array}\right\} \tag{3.5}$$

Integration in eqn. 3.4 is over the x and y dimensions of the aperture only. Clearly for separable aperture distributions calculation of the radiation pattern reduces to the product of two two-dimensional patterns. We choose as a simple example a rectangular aperture with uniform amplitude and phase and normalise this electric field distribution to unity. Then $E(x,y) = 1$ in eqn. 3.1 and with the aperture dimensions as in Fig. 3.1, the field is given by eqn. 3.3 with

$$F_1(k_1) = \int_{-a/2}^{a/2} \exp(jk_1 x) dx = 2 \int_0^{a/2} \cos(k_1 x) dx = a \frac{\sin(k_1 a/2)}{(k_1 a/2)}$$

and

$$\tag{3.6}$$

$$F_2(k_2) = b \frac{\sin(k_2 b/2)}{(k_2 b/2)} \tag{3.7}$$

This gives the three-dimensional radiation pattern in $-\pi/2 < \theta < \pi/2$. For practical reasons it is usually the patterns in the two principal planes, here the x–z ($\phi = 0$) and y–z ($\phi = \pi/2$) planes, which are of most interest. These are, respectively,

$$\bar{E}(r,\theta,0) = \frac{\hat{\theta} j \exp(-jkr)}{2\lambda r}(1 + \cos\theta) ab \frac{\sin[(\pi a/\lambda)\sin\theta]}{(\pi a/\lambda)\sin\theta} \tag{3.8}$$

and

$$\bar{E}(r,\theta,\pi/2) = -\hat{\phi} \frac{j \exp(-jkr)}{2\lambda r}(1 + \cos\theta) ab \frac{\sin[(\pi b/\lambda)\sin\theta]}{(\pi b/\lambda)\sin\theta} \tag{3.9}$$

Both principal plane patterns are of the same form but scaled in θ according to the aperture dimensions in the respective planes. The following considerations of the pattern in the x–z plane accordingly apply in the y–z plane with the aperture dimension b replacing a.

The radiation pattern is represented by the function $\sin x/x$ with $x = (\pi a/\lambda) \sin\theta$ in Fig. 3.2. Nulls in the pattern occur when $(\pi a/\lambda)\sin\theta = n\pi, n = \pm 1, \pm 2, \ldots$, so the full angular width of the main beam, from the null at $n = -1$ to that at $n = 1$, is $2 \sin^{-1}(\lambda/a) \approx 2\lambda/a$ radians if $a \gg \lambda$. The half-power point on the main lobe, where the field is down to $\sqrt{0\cdot5} = 0\cdot707$ of its peak value, is at about $(\pi a/\lambda)\sin\theta = 1\cdot39$. Hence the half-power or 3 dB beamwidth is

$$\theta_{1/2} = 2 \sin^{-1}\left(\frac{1 \cdot 39\lambda}{\pi a}\right) \tag{3.10}$$

radians or, for large apertures $(a \gg \lambda)$, $0 \cdot 88 \ \lambda/a$ radians or $50 \cdot 7 \ \lambda/a$ degrees. The first sidelobes are at $(\pi a/\lambda) \sin \theta = \pm 1 \cdot 43\pi$ and are $20 \log_{10} (0 \cdot 217) = -13 \cdot 3$ dB below the peak value of the main beam. Radiation patterns are usually characterised by their half-power beamwidth and the level of the first sidelobes.

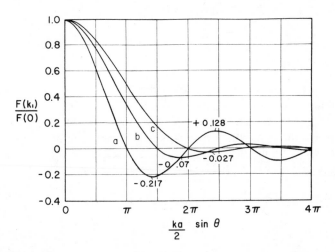

Fig. 3.2 *Radiation patterns of symmetrical field distributions in a rectangular aperture*
Aperture distributions:
(a) uniform
(b) cosinusoidal
(c) cosine-squared

It is apparent from these results that the larger the dimensions of an aperture in wavelengths, the smaller the beamwidth of the pattern, a result valid for all in-phase distributions.

3.2 Simple aperture distributions

Table 3.1 gives the radiation pattern for several simple aperture distributions and, where applicable, the beamwidth and first sidelobe level. These distributions can be represented as either even or odd functions about the origin of coordinates at the aperture centre. The radiation pattern in the corresponding principal plane is then

$$F(k_1) = \int_{-a/2}^{a/2} E(x) \exp(jk_1 x)\,dx = 2 \int_{0}^{a/2} E(x) \cos(k_1 x)\,dx,$$
$$E(x) \text{ even}$$
$$= 2j \int_{0}^{a/2} E(x) \sin(k_1 x)\,dx, \tag{3.11}$$
$$E(x) \text{ odd}$$

with $k_1 = \mathrm{k} \sin \theta$, for a rectangular aperture dimension a.

Table 3.1 *Radiation patterns of simple aperture distributions*

Aperture distribution	$E(x)$	$F(k_1)$	Full- and half-power beamwidths (radians)	First sidelobe level
cosine	$\cos\left(\dfrac{\pi x}{a}\right)$	$2\pi a\,\dfrac{\cos(k_1 a/2)}{\pi^2 - (k_1 a)^2}$	$2\sin^{-1}(1\cdot5\lambda/a)$ $\theta_{1/2} \approx 1\cdot2\lambda/a$	$0\cdot070$ $(-23\cdot1\text{ dB})$
cosine-squared	$\cos^2\left(\dfrac{\pi x}{a}\right)$	$\dfrac{a}{2}\cdot\dfrac{\sin(k_1 a/2)}{k_1 a/2}\,\dfrac{\pi^2}{\pi^2 - (k_1 a/2)^2}$	$2\sin^{-1}(2\lambda/a)$ $\theta_{1/2} \approx 1\cdot44\lambda/a$	$0\cdot026$ $(-31\cdot5\text{ dB})$
antiphase constant	$1;\dfrac{a}{2} > x > 0$ $-1; 0 > x > -\dfrac{a}{2}$	$\dfrac{a}{}\,\dfrac{1 - \cos(k_1 a/2)}{(k_1 a/2)}$	These patterns have nulls at $k_1 = 0$	
sine	$\sin\left(\dfrac{2\pi x}{a}\right)$	$-ja\,\dfrac{\pi\sin(k_1 a/2)}{\pi^2 - (k_1 a/2)^2}$		

Aperture distributions which are even functions, i.e. symmetrical about the aperture centre, produce patterns symmetrical about $\theta = 0$. Patterns of three symmetrical distributions are shown in Fig. 3.2 normalised to the peak field on the beam-axis. The uniform distribution has the narrowest beam and highest sidelobe levels of the three. Later it will be shown to provide the highest gain of all in-phase distributions; in this sense it is an ideal distribution. In most real distributions the field decreases towards the aperture edges, resulting in broader beams and lower sidelobe levels.

Aperture distributions which are odd functions yield patterns which are odd functions, i.e. there is a null at $\theta = 0$, as indicated in Table 3.1. Such patterns may be useful for antennas in guiding or tracking on a pattern null, but otherwise are rarely used. Similar distributions may also appear as the cross-polarised fields of a paraboloidal reflector. Patterns of aperture distributions which are odd functions and other illustrations of the application of Fourier transforms to antenna theory are in a series of articles by Ramsay (1946–47).

3.3 Compound aperture distributions

The aperture field distribution $E(x)$ and the radiation pattern $F(k_1)$ are related by

$$F(k_1) = \int_{-\infty}^{\infty} E(x) \exp(jk_1 x) dx$$

$$E(x) = \frac{1}{2\pi} \int_{-\infty}^{\infty} F(k_1) \exp(-jk_1 x) dk_1$$

(3.12)

The radiation pattern of the aperture distribution $E_1(x) + E_2(x)$ is

$$\int_{-\infty}^{\infty} [E_1(x) + E_2(x)] \exp(jk_1 x) dx = F_1(k_1) + F_2(k_1)$$

where $F_1(k_1)$ is the pattern of $E_1(x)$ and $F_2(k_1)$ of $E_2(x)$. In general, the radiation pattern of the distribution

$$\sum_{n=1}^{m} C_n E_n(x) \text{ will be } \sum_{n=1}^{m} C_n F_n(k_1)$$

(3.13)

where the coefficients C_n are independent of x and k_1.

This means the radiation patterns for compound aperture distributions may be got by adding suitably weighted patterns for simple distributions. For example, the aperture distribution

$$E(x) = C + (1 - C) \cos^2 \left(\frac{\pi x}{a} \right), \quad |x| < a/2$$

(3.14)

$$= 0 \qquad\qquad\quad , \quad |x| > a/2$$

produces the radiation pattern

$$F(k_1) = a \frac{\sin(k_1 a/2)}{k_1 a/2} \left[C + \frac{(1-C)}{2} \frac{\pi^2}{\pi^2 - (k_1 a/2)^2} \right] \tag{3.15}$$

This is illustrated in Fig. 3.3 for $C = 1/3$, where the first null in eqn 3.15 occurs when $k_1 a/2 = \sqrt{2}\,\pi$. The full beamwidth is then $2\sin^{-1}(\sqrt{2}\lambda/a)$ radians $\approx 2\sqrt{2}\lambda/a$, $a \gg \lambda$, compared to $2\lambda/a$ for the uniform distribution and $4\lambda/a$ for the cosine-squared distribution. This pattern is a reasonably efficient compromise between the pattern

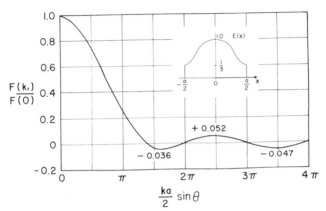

Fig. 3.3   *Compound aperture distribution and its radiation pattern*

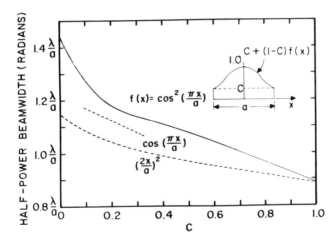

Fig. 3.4   *Half-power beamwidths of some compound symmetrical aperture distributions*

for a uniform distribution, which has a narrow beam but high sidelobe levels, and the cosine-squared pattern with its low sidelobe levels but broad beam. A distribution similar to Fig. 3.3 might appear in the aperture of a parabolic reflector antenna. Then the aperture edge illumination can be altered through the feed directivity to provide the secondary pattern desired. Fig. 3.4 shows the pattern half-power beamwidth for the aperture distribution given by eqn. 3.15 as well as for other distributions.

The effect of aperture blockage by a reflector feed or subreflector also can be investigated by the present theory. Fig. 3.5a shows a typical field distribution in the aperture of an antenna reflector of width a with a feed of width $\delta \ll a$ in the centre of the aperture plane. Without blocking the aperture distribution is $E(x)$ and the pattern $F(k_1)$. With aperture blocking the pattern is

$$F'(k_1) = \left(\int_{-a/2}^{-\delta/2} + \int_{\delta/2}^{a/2} = \int_{-a/2}^{a/2} - \int_{-\delta/2}^{\delta/2} \right) E(x)\exp(jk_1 x)\,dx$$

$$\simeq F(k_1) - \frac{\delta \sin(k_1 \delta/2)}{k_1 \delta/2}$$

since $E(x) \simeq 1$ for $-\delta/2 < x < \delta/2$. The second term in the last equation is the radiation pattern for a uniform aperture distribution of width δ. Since δ is small

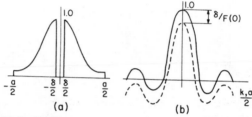

Fig. 3.5   *Typical field distributions*
(a) Blocked aperture distribution
(b) Effect on the radiation pattern. Unblocked pattern _____$F(k_1)//F(0)$; blocked pattern - - - - -$F'(k_1)//F(0)$.

the pattern is broad so that over the central portion of the antenna radiation pattern $F'(k_1) \simeq F(k_1) - \delta$. The patterns $F(k_1)/F(0)$ and $F'(k_1)/F(0)$ are shown in Fig. 3.5b, where it is apparent that aperture blockage narrows the main beamwidth. It also raises the level of the first sidelobe. If $F(0) = 1$ and p is the normalised level of the first sidelobe in the unblocked pattern, then $F'(0)/F(0) = 1 - \delta/F(0)$ and $F'/F(0) = p + \delta/F(0)$ at the first sidelobe in the blocked pattern. The sidelobe level in the blocked pattern is

$$p' = \frac{p + \delta/F(0)}{1 - \delta/F(0)} \tag{3.16}$$

which always exceeds p.

3.4 Displaced aperture distributions

Assuming eqn. 3.12, let $E_1(x) = E(x - x_0)$, which implies a lateral shift x_0 of the distribution in the aperture plane. The pattern is

$$F_1(k_1) = \int_{-\infty}^{\infty} E(x - x_0)\exp(jk_1 x)\,dx = \int_{-\infty}^{\infty} E(\xi)\exp[jk_1(\xi + x_0)]\,d\xi$$

$$= F(k_1)\exp(jk_1 x_0) \tag{3.17}$$

The effect on the radiation pattern is a linear phase change of $k_1 x_0$ radians. Usually in measuring a pattern field strength magnitude is observed and there is no apparent change in the pattern at a range $r \gg x_0$. Hence in far-field antenna pattern measurements there is no need for the centre of rotation to coincide with the centre of the aperture under test.

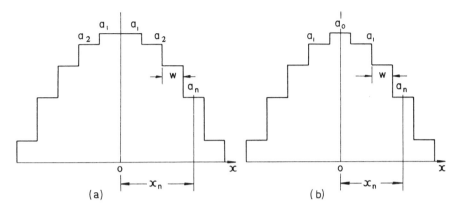

Fig.3.6 *Arrays of (a) even and (b) odd numbers of uniform apertures with amplitudes symmetrical about the array centre*

We may use eqn. 3.17 together with eqn. 3.13 to obtain the total pattern of a group of m apertures, at distances x_n from the centre of coordinates in the aperture plane. The pattern of

$$\sum_{n=1}^{m} a_n E_n(x - x_n) \text{ is } \sum_{n=1}^{m} a_n F_n(k_1) \exp(jk_1 x_n)$$

As an example Fig. 3.6a shows an array of $2m$ uniform in-phase apertures of width w placed end-to-end and with amplitudes a_n symmetrical about $x = 0$. The pattern is

$$F(k_1) = w \frac{\sin(k_1 w/2)}{k_1 w/2} \sum_{n=1}^{m} a_n \{\exp[j(2n-1)(k_1 w/2)]$$

$$+ \exp[-j(2n-1)(k_1 w/2)]\}$$

$$= 2w \frac{\sin(k_1 w/2)}{k_1 w/2} \sum_{n=1}^{m} a_n \cos[(2n-1)(k_1 w/2)] \tag{3.18}$$

and for an odd number $2m + 1$ such apertures with a symmetrical in-phase distribution (Fig. 3.6b)

$$F(k_1) = w \frac{\sin(k_1 w/2)}{k_1 w/2} \left[a_0 + 2 \sum_{n=1}^{m} a_n \cos(nk_1 w) \right] \tag{3.19}$$

In this way patterns of arbitrary aperture distributions can be obtained by summing contributions from segments of the aperture of equal width and essentially uniform amplitude and phase.

The expressions eqns. 3.18 and 3.19 will be recognised as the product of element and group patterns of an antenna array. They could represent, for example, the E-plane pattern of a stack of in-phase large rectangular waveguides supporting only the TE_{10} mode. For the H-plane pattern of an array of such waveguides side-by-side the appropriate element pattern would be that for the cosinusoidal distribution, given in Table 3.1.

3.5 Aperture fields with linear phase variations

If $F(k_1)$ is the Fourier transform of $E(x)$, the transform of $E(x) \exp(-jk_0 x)$, the same distribution but with a linear phase variation is

$$\int_{-\infty}^{\infty} E(x) \exp(-jk_0 x + jk_1 x) dx = F(k_1 - k_0) \qquad (3.20)$$

or $F(k \sin \theta - k_0)$. Hence a linear phase variation $k_0 x$ across an aperture produces an angular displacement $\theta_0 = \sin^{-1}(k_0/k)$ in the radiation pattern. This may be used as the basis of an antenna scanning technique. For example, if an aperture field phase variation of $\exp(-jks \sin \theta_0)$ is introduced to a uniform constant phase distribution the pattern becomes

$$F[k(\sin \theta - \sin \theta_0)] = a \frac{\sin[(\pi a/\lambda)(\sin \theta - \sin \theta_0)]}{(\pi a/\lambda)(\sin \theta - \sin \theta_0)} \qquad (3.21)$$

which has its maximum at $\theta = \theta_0$. There is also an increase in the width of the main beam by a factor inversely proportional to the projected width of the aperture in the direction θ_0, i.e. by $(\cos \theta_0)^{-1}$.

We may also use eqn. 3.20 in an antenna pattern synthesis technique (Woodward, 1946). The aperture distribution

$$E(x) = \sum_{n=-m}^{m} b_n \exp[-j2\pi(nx/a)] \qquad (3.22)$$

consists of $2m + 1$ distributions of uniform amplitude b_n with a linear phase variation of $2\pi n$ radians across the aperture. The radiation pattern will be a linear superposition of shifted functions of the form eqn. 3.21:

$$F(k_1) = a \sum_{n=-m}^{m} b_n \frac{\sin[(\pi a/\lambda) \sin \theta - n\pi]}{(\pi a/\lambda) \sin \theta - n\pi} \qquad (3.23)$$

Each term in the sum of eqn. 3.23 has a peak value of b_n at $\theta = \sin^{-1}(n\lambda/a)$. If the coefficients b_n are set equal to a desired pattern $F_1(k_1)$ at these angles, i.e. if $b_n = F_1(2\pi n/a)$, eqn. 3.23 will approximate it. The accuracy of this approximation will be limited by the aperture width a, since the largest m for real angles θ is a/λ. This limitation can be overcome (Woodward and Lawson, 1948) but usually at the expense of 'supergain' effects discussed later. The same technique is used in the sampled data representation of band-limited signals.

3.6 Gaussian distribution

The transform of the Gaussian function $E(x) = \exp(-x^2)$,

$$
\begin{aligned}
F(k_1) &= \int_{-\infty}^{\infty} \exp(-x^2 + jk_1 x)\,dx \\
&= \exp\!\left(\frac{-k_1^2}{4}\right) \int_{-\infty}^{\infty} \exp\{-[x - j(k_1/2)]^2\}\,dx = \sqrt{\pi}\,\exp(-k_1^2/4)
\end{aligned}
$$
(3.24)

is itself a Gaussian function. Thus a Gaussian distribution in an aperture of infinite extent gives a radiation pattern with no sidelobes.

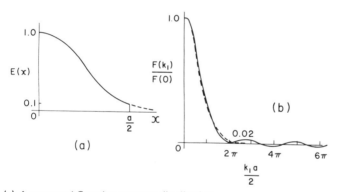

Fig. 3.7   (a) A truncated Gaussian aperture distribution
(b) Its radiation pattern
The dashed curve in (a) is $\exp[-2\cdot3(2x/a)^2]$ and in (b) its Fourier transform is $\exp[-0\cdot435(k_1 a/2)^2]$

In practice, the Gaussian aperture distribution is truncated at the aperture edges. Then the radiation pattern is

$$
\begin{aligned}
F(k_1) &= \int_{-a/2}^{a/2} \exp(-\alpha x^2 + jk_1 x)\,dx \\
&= \exp(-k_1^2/4\alpha)\left(\int_0^{a/2} - \int_0^{-a/2}\right)\exp\{-\alpha[x - j(k_1/2\alpha)]^2\}\,dx \\
&= \exp(-k_1^2/4\alpha)\,\frac{1}{2}\sqrt{\frac{\pi}{\alpha}}\left[\operatorname{erf}\!\left(\frac{a}{2}\sqrt{\alpha} - j\frac{k_1}{2\sqrt{\alpha}}\right)\right. \\
&\quad \left. - \operatorname{erf}\!\left(-\frac{a}{2}\sqrt{\alpha} - j\frac{k_1}{2\sqrt{\alpha}}\right)\right]
\end{aligned}
$$
(3.25)

where

$$\text{erf}(u) = \frac{2}{\sqrt{\pi}} \int_0^u \exp(-t^2)\,dt \tag{3.26}$$

is the error function, which has been tabulated for complex arguments (Faddeyeva and Terent'ev, 1961). Suppose $E(\pm a/2) = 0\cdot1$, i.e. the amplitude taper is $\alpha = -(4/a^2)\ln(0\cdot1) = 9\cdot21/a^2$. Then the pattern is shown in Fig. 3.7(b). The half-power beamwidth is $2\sin^{-1}(0\cdot86\lambda/a)$ and the normalised first sidelobe level is about $0\cdot02$. The sidelobe envelope decreases as $\sin(k_1 a/2)/(k_1 a/2)$.

The corresponding odd distribution function, $E(x) = xe^{-x^2}$ is, of course, also self-reciprocal.

3.7 Circular aperture

Fig. 3.8 shows a circular aperture of diameter a in the $z = 0$ plane. If the aperture

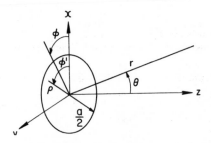

Fig. 3.8 *Coordinates for radiation from a circular aperture*

is large ($ka \gg 1$) and the electric field in the aperture $E_x(\rho',\phi')$ is polarised in the x-direction, its far field is given by eqn. 3.1. With $x = \rho'\cos\phi'$, $y = \rho'\sin\phi'$, this becomes

$$\bar{E}(r,\theta,\phi) = \bar{A} \int_0^{2\pi} \int_0^{a/2} E_x(\rho',\phi') \exp[jk\rho'\sin\theta\cos(\phi-\phi')]\,\rho'd\rho'd\phi'$$

$$= \bar{A}\frac{a^2}{4} \int_0^{2\pi} \int_0^1 E_x\left(\frac{la}{2},\phi'\right) \exp[jk_1(la/2)\cos(\phi-\phi')]\,ldld\phi' \tag{3.27}$$

where $k_1 = k\sin\theta$ and $l = 2\rho'/a$. If the aperture field is uniform in amplitude and phase, i.e. $E_x[(a/2),\phi'] = 1$, the integral form of the Bessel function J_0 replaces integration in ϕ' and

$$\bar{E}(r,\theta,\phi) = \bar{A}\frac{\pi a^2}{2} \int_0^1 J_0\left(\frac{k_1 al}{2}\right) l\,dl$$

$$= \bar{A}\frac{\pi a^2}{2} \frac{J_1(k_1 a/2)}{(k_1 a/2)} \tag{3.28}$$

This pattern shown in Fig. 3.9 ($n = 0$) is very similar to the principal plane patterns of eqns. 3.8 and 3.9 but the half-power beamwidth

$$\theta_{1/2} = 2 \sin^{-1} \left(0{\cdot}51 \frac{\lambda}{a} \right) \tag{3.29}$$

is larger than eqn. 3.10. The first sidelobe is $17{\cdot}5$dB down from the peak value, lower than for a uniform distribution in a rectangular aperture.

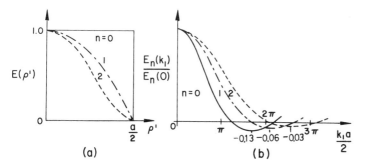

Fig. 3.9 *a* Aperture distributions
b Radiation patterns of a circular aperture of diameter *a*

Non-uniform aperture illuminations yield patterns with broader beams and lower sidelobes. In eqn. 3.27, let

$$E(l,\phi') = (1 - l^2)^n, \quad n = 0, 1, 2, \ldots$$

Then

$$\bar{E}_n(r,\theta,\phi) = \bar{A} \frac{\pi a^2}{2} \int_0^1 (1 - l^2)^n J_0\left(\frac{k_1 a l}{2} \right) l dl$$

$$= \bar{A} \frac{\pi a^2}{4(n + 1)} \Lambda_{n+1} \left(\frac{k_1 a}{2} \right) \tag{3.30}$$

where Λ_n are tabulated functions (Jahnke and Emde, 1945, p. 180). Normalised patterns for $n = 0, 1$ and 2 are shown in Fig. 3.9.

3.8 Operational calculation of radiation patterns

Differentiation of

$$F(k_1) = \int_{-\infty}^{\infty} E(x) \exp(jk_1 x) dx$$

n times with respect to k_1 gives

$$\frac{\partial^n F(k_1)}{\partial k_1^n} = \int_{-\infty}^{\infty} (jx)^n E(x) \exp(jk_1 x) dx \tag{3.31}$$

Dividing by j^n, the pattern resulting from the distribution

$$x^n E(x) \text{ is } (-jD)^n F(k_1)$$

where the operator $D^n = \partial^n / \partial k_1{}^n$. Similarly, the pattern of

$$\sum a_n x^n E(x) \quad \text{is} \quad \sum a_n (-jD)^n F(k_1)$$

or, since any continuous function $f(x)$ can be represented by a polynomial, the pattern of

$$f(x)E(x) \quad \text{is} \quad f(-jD)F(k_1). \tag{3.32}$$

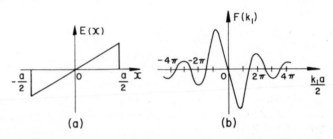

Fig. 3.10 *a* Rectilinear aperture distribution
  b Its radiation pattern

As an example, the radiation pattern of the rectilinear distribution $E(x) = 2x/a$ can be obtained by differentiation of the pattern of the uniform distribution. The pattern is

$$\frac{2}{a}\left[-j\frac{\partial}{\partial k_1}\left(a\,\frac{\sin(k_1 a/2)}{(k_1 a/2)}\right)\right] = -j\,\frac{a}{(k_1 a/2)^2}\left[\frac{k_1 a}{2}\cos\left(\frac{k_1 a}{2}\right) - \sin\left(\frac{k_1 a}{2}\right)\right]$$

which is illustrated in Fig. 3.10. Another example of this operational method is in Section 4.3.

Chapter 4

Near-field radiation patterns

4.1 Near-field criteria

In the previous chapter, the radiation patterns considered were at distances from the aperture sufficiently large that all ray paths from the aperture to the field point are essentially parallel, i.e. $r \gg a,b$ in Fig. 3.1 Here this assumption is removed and the radiation field in what is known in optics as the Fresnel zone of the aperture is examined. With the condition $r \gg \lambda$ maintained, the reactive near fields of the aperture are excluded from consideration.

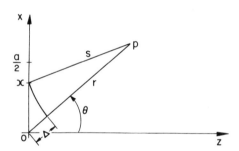

Fig. 4.1 *Path-length differences in the near field of an aperture*

Fig. 4.1 shows ray paths xp and op in the $x-z$ plane which are not parallel. From the geometry

$$s^2 - r^2 = (s-r)(s+r) = x^2 - 2rx \sin \theta.$$

If Δ is the path difference at p of radiation from o and x in the aperture $(s-r) = -\Delta$ and, since $s+r \simeq 2r$,

$$\Delta \simeq \frac{-x^2}{2r} + x \sin \theta. \tag{4.1}$$

The effect of a finite range on the path difference is the quadratic term in eqn 4.1 which vanishes as $r \to \infty$. The usual requirement for the far-field patterns is that the maximum error in path length due to finite range is less than $\lambda/16$ for radiation from all parts of the aperture. For an aperture of width a the criterion is $a^2/8r < \lambda/16$ or

$$r > \frac{2a^2}{\lambda} \qquad (4.2)$$

This is the rather arbitrary division between the near field or Fresnel zone of an aperture and the far field. The effect of the finite range on the radiation pattern will depend also on the aperture distribution. The patterns of distributions highly tapered towards the aperture edges will be less affected by a particular finite range than the patterns of more uniform distributions. In consequence, this criterion may be inadequate in the latter case and more than adequate in the former. Near-field effects are also larger for aperture distributions tapered in phase towards aperture edges than for in-phase distributions.

The importance of near-field effects is readily indicated by an example. A typical heavy route satellite station antenna has a paraboloidal reflector diameter of 30 m. At an up-frequency of 6 GHz, the minimum range for radiation pattern measurements is 36 km, according to eqn. 4.2. Consequently, any conventional pattern measurements are necessarily near-field and the far-field pattern is measured by letting a satellite pass through the beam. In most other near-field situations there is no such simple solution and the aperture need not be physically large for the problem to arise. A 1 m diameter paraboloidal reflector at 35 GHz, for example, requires a test range in excess of 230 m. A range of this length free of multipath interference is difficult to arrange and often one must settle for a pattern measured in the near-field.

4.2 Near-field patterns from aperture distributions

Eqn. 4.1 shows that the phase difference between contributing waves is $2\pi\Delta/\lambda = -\beta x^2 + k_1 x$ where $\beta = \pi/r\lambda$ and $k_1 = k \sin \theta$. If the aperture field distribution is $E_1(x)$ the near-field radiation pattern is

$$D_1(k_1) = \int_{-\infty}^{\infty} E_1(x) \exp(-j\beta x^2) \exp(jk_1 x) \, dx \qquad (4.3)$$

which in optics is called a Fresnel transform.

A comparison of eqns 4.3 and 3.4 shows the first-order effect of a finite range can be simulated by a quadratic phase variation in the aperture distribution. Similar results are obtained for the pattern in the $y-z$ plane of Fig. 3.1 and the near-field radiation pattern is

$$\bar{E}(r,\theta,\phi) = \bar{A}D_1(k_1)D_2(k_2), \qquad (4.4)$$

where

$$D_1(k_1) = \int_{-a/2}^{a/2} E_1(x) \exp[-j(\beta x^2 - k_1 x)] \, dx$$

$$D_2(k_2) = \int_{-b/2}^{b/2} E_2(y) \exp[-j(\beta y^2 - k_2 y)] \, dy$$

(4.5)

if $E_1(x) = 0$, $|x| > a/2$ and $E_2(y) = 0$, $|y| > b/2$. In eqn. 4.5, k_1 and k_2 are defined by eqn. 3.5. Clearly as $r \to \infty, \beta \to 0$ and eqn. 4.5 becomes eqn. 3.4.

The integrals of eqn. 4.5 are more involved than those for the far-field radiation pattern but for simple aperture distributions they can be reduced to Fresnel integrals. For example, the uniform distribution $E_1(x) = 1$ in eqn. 4.5 gives

$$D_1(k_1) = \exp[j(k_1^2/4\beta)] \int_{-a/2}^{a/2} \exp[-j\beta(x - k_1/2\beta)^2] \, dx$$

$$= \sqrt{\frac{r\lambda}{2}} \exp[j(r\lambda k_1^2/4\pi)] \int_{u_1}^{u_2} \exp[-j(\pi/2)u^2] \, du$$

$$= \sqrt{\frac{r\lambda}{2}} \exp[j(r\lambda k_1^2/4\pi)] \{C(u_2) - C(u_1) - j[S(u_2) - S(u_1)]\}$$

(4.6)

where

$$\begin{matrix} u_2 \\ u_1 \end{matrix} = \pm \frac{a}{\sqrt{2r\lambda}} - \frac{k_1}{2\pi} \sqrt{2r\lambda}$$

(4.7)

and the Fresnel integrals defined by

$$\int_0^{v_1} \exp[-j(\pi/2)v^2] \, dv = C(v_1) - jS(v_1)$$

(4.8)

are tabulated (Jahnke and Emde, 1945; Abramowitz and Stegun, 1964). Computer subroutines are also usually available.

The near-field pattern for a cosinusoidal distribution can be treated similarly. If, in eqn. 4.5,

$$E_2(y) = \cos\frac{\pi y}{b} = \frac{1}{2}\{\exp[j(\pi y/b)] + \exp[-j(\pi y/b)]\}$$

then

$$D_2(k_2) = \frac{1}{2}\left(\exp\left[j\frac{1}{4\beta}\left(k_2 + \frac{\pi}{b}\right)^2\right] \int_{-b/2}^{b/2} \exp\left\{-j\beta\left[y - \frac{1}{2\beta}\left(k_2 + \frac{\pi}{b}\right)\right]^2\right\}\right.$$

$$\times \, dy + \exp\left[j\frac{1}{4\beta}\left(k_2 - \frac{\pi}{b}\right)\right]$$

$$\left. \int_{-b/2}^{b/2} \exp\left\{-j\beta\left[y - \frac{1}{2\beta}\left(k_2 - \frac{\pi}{b}\right)^2\right]\right\}\right\} dy\right)$$

$$= \frac{1}{2}\sqrt{\frac{r\lambda}{2}}\left(\exp\left[j\frac{r\lambda}{4\pi}\left(k_2 + \frac{\pi}{b}\right)^2\right] \int_{v_1}^{v_2} \exp[-j(\pi v^2/2)] \, dv\right.$$

$$\left. + \exp\left[j\frac{r\lambda}{4\pi}\left(k_2 - \frac{\pi}{b}\right)^2\right] \int_{w_1}^{w_2} \exp[-j(\pi w^2/2)] \, dw\right)$$

$$= \frac{1}{2} \sqrt{\frac{r\lambda}{2}} \left(\exp\left[j \frac{r\lambda}{4\pi} \left(k_2 + \frac{\pi}{b} \right)^2 \right] \right.$$

$$\times \{ C(v_2) - C(v_1) - j[S(v_2) - S(v_1)] \}$$

$$+ \exp\left[j \frac{r\lambda}{4\pi} \left(k_2 - \frac{\pi}{b} \right)^2 \right]$$

$$\left. \times \{ C(w_2) - C(w_1) - j[S(w_2) - S(w_1)] \} \right) \tag{4.9}$$

where

$$\left. \begin{aligned} v_{\substack{2\\1}} &= \pm \sqrt{\frac{b}{2r\lambda}} - \frac{k_2}{2\pi} \sqrt{2r\lambda} - \frac{1}{b}\sqrt{\frac{r\lambda}{2}} \\ w_{\substack{2\\1}} &= \pm \sqrt{\frac{b}{2r\lambda}} - \frac{k_2}{2\pi} \sqrt{2r\lambda} + \frac{1}{b}\sqrt{\frac{r\lambda}{2}} \end{aligned} \right\} \tag{4.10}$$

Aperture distributions which are sinusoidal or cosinusoidal functions can be dealt with in this way, by expressing the entire integrand as a complex exponential and completing the square to reduce the expression to Fresnel integrals. Compound distributions can be decomposed into simple distributions and so treated also, but direct numerical integration will often be simpler.

4.3 Near-field patterns from far-field patterns

If the radiation pattern expression is an analytic function its near-field pattern can be calculated from it and its even-order derivatives. The operational method, specifically eqn. 3.32, is used to obtain the radiation pattern of

$$\exp(-j\beta x^2)E(x) \text{ as } \exp[-j\beta(-jD)^2]F(k_1) = \exp(j\beta D^2)F(k_1)$$

where $D = \partial/\partial k_1$ and $\beta = \pi/r\lambda$. Since $\beta \ll 1$ for this representation to be valid, only the first few terms of the power series expansion of the operator

$$\exp(j\beta D^2) = 1 + j\beta D^2 - \frac{\beta^2}{2!} D^4 - j \frac{\beta^3 D^6}{3!} + \ldots$$

are retained giving

$$D(k_1) = F(k_1) + j\beta \frac{\partial^2}{\partial k_1^2} F(k_1) - \frac{\beta^2}{2!} \frac{\partial^4 F(k_1)}{\partial k_1^4} - j \frac{\beta^3}{3!} \frac{\partial^6 F(k_1)}{\partial k_1^6} + \ldots \tag{4.11}$$

Tabulated values of the derivatives of the Fourier transforms of uniform and cosinusoidal distributions are available along with other data for the use of this method (Spencer and Austin, 1946; Milne, 1952).

The near-field radiation patterns of a cosinusoidal distribution shown in Fig. 4.2 were obtained in this way. The effects of a finite range evident are typical; there is a broadening of the main beam, a filling-in of the pattern nulls and a raising of the sidelobe levels, all of which become more pronounced as the range diminishes. These effects are also evident in the near-field patterns of circular aperture antennas (Hansen and Bailin, 1959).

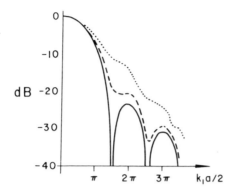

Fig. 4.2 *Near-field radiation patterns in the $\phi = \pi/2$ plane of the distribution $E_2 (y) = \cos$ $(\pi y /b)$ in the aperture of Fig. 3.1.*
Ranges: $\cdots\cdots r = b^2 /3\lambda; \text{- - - - -} r = b^2 /\lambda; \text{———} r = \infty$

4.4 Effect of measuring antenna directivity

The observed radiation pattern is unaffected by either the range or the directivity of the measuring antenna in the far field, but not in the near field. Unless the measuring antenna is essentially omnidirectional the pattern observed in the near field is a selection of the incident spectrum of plane waves weighted in accordance with the pattern of the receiving antenna.

Using the plane wave spectrum concepts of Chapter 2 to introduce the pattern $F^R(k_x',k_y')$ of the measuring antenna Brown (1958a, b) obtained the following expression for the near-field response of the receiving antenna to the radiation field of eqn. 2.36 in the $y-z$ ($\phi = \pi/2$) plane of Fig. 3.1.

$$D = \frac{Y_0}{8\pi^2} \int_{-\infty}^{\infty} \int_{-\infty}^{\infty} (k^2 - k_y'^2)F(k_x,k_y)F^R(k_x',k_y') \exp(-jk_z'r)$$

$$\frac{dk_x' dk_y'}{k_z'} \tag{4.12}$$

where $k_x' = k \sin \alpha' \cos \beta'$, $k_y' = k \sin \alpha' \sin \beta'$ and $k_z' = k \cos \alpha'$ and α' and β' are angles referred to the coordinates of the receiving antenna of Fig. 4.3. The radiated field of eqn. 2.36 has no E_y component and the received field of eqn. 4.12 rejects the E_y' component of incident field. No approximations were made in deriving eqn. 4.12 but its evaluation is in general very difficult. In the far field, $kr \gg 1$, integration by stationary phase is possible, yielding for $\phi = \pi/2$.

$$D \simeq j \, \frac{Y_0 \cos \theta}{2\lambda r} \, \exp(-jkr)F(0, k \sin \theta)F^R(0,0) \tag{4.13}$$

which is independent of the radiation pattern of the receiving antenna.

A near-field evaluation of eqn. 4.12 can be made for antennas large in the b-dimension of Fig. 3.1 but relatively small in the a-dimension. The measuring antenna is in the far field of the smaller aperture dimension but in the near field

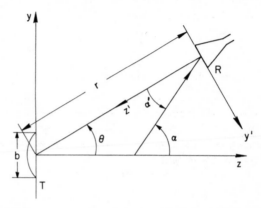

Fig. 4.3   *Arrangement for measurement of the $\phi = \pi/2$ plane radiation pattern of antenna T with antenna R*

of the larger and directed at the measured antenna. With radiation incident upon it at angles $\alpha' = \alpha - \theta$ (see Fig. 4.3) mainly over the central portion of its main beam a good approximation to its pattern in the $\beta' = 0$ plane is (Brown and Jull, 1961)

$$F^R[0, k \sin \alpha'] = F^R(0,0) \exp[-K(1 - \cos \alpha')] \tag{4.14}$$

where

$$K = \frac{\ln 2}{2\{1 - (\cos B_R/2)\}} \tag{4.15}$$

and B_R is the half-power beamwidth of the measuring antenna. This pattern has the beamwidth of the receiving antenna and for small α' is essentially Gaussian. With eqn. 4.14 in eqn. 4.12, an asymptotic expansion for the observed near-field pattern is

$$D(\theta) \simeq \frac{Y_0}{2\lambda}\left(\frac{jk}{\rho r}\right)^{1/2} \exp(-jkr) \sum_{n=0}^{m} a_n \frac{\partial^{2n}}{\partial \theta^{2n}} F(k_1) \tag{4.16}$$

where $k_1 = k \sin \theta$,

$$a_0 = 1 + \frac{j}{8\rho} - \frac{9}{128\rho^2} - j \frac{75}{1024\rho^3} + \ldots$$

$$a_1 = \frac{j}{2\rho} - \frac{5}{16\rho^2} - j \frac{259}{768\rho^3} + \ldots$$

$$a_2 = -\frac{1}{8\rho^2} - j\frac{35}{192\rho^3} + \ldots$$
(4.17)

$$a_3 = -\frac{j}{48\rho^3} + \ldots$$

and

$$\rho = kr + jK$$
(4.18)

The first terms in the series expansion of the coefficients a_n was given in the original derivation by Brown (1958a), i.e.

$$a_n = \frac{1, 3 \ldots (2n-1)}{(2n)! \, (-j\rho)^n}, \; n = 1, 2, \ldots$$
(4.19)

and the higher-order terms were obtained later (Jull, 1963). The additional terms associated with the lower order derivatives of $F(k_1)$ are required for pattern calculations at the shorter near field ranges. Only a few terms in the series of eqn. 4.16 are needed. For large B_R the series reduces to that of eqn. 4.11.

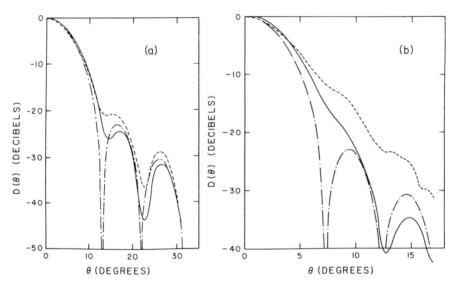

Fig. 4.4 *Predicted effects of finite range and directive measuring antennas on the pattern of the distribution* $E_2(y) = \cos(\pi y/b)$.
—·— far-field patterns
a Range $r = b^2/\lambda$. Measuring antenna aperture width: - - - - - 0·87λ; ——— 10·3λ
b Range $r = 0·36b^2/\lambda$. Measuring antenna aperture width: - - - - - 0·98λ; ——— 11·7λ

Some numerical results based on eqn. 4.16 are given in Figs. 4.4*a* and 4.4*b*. These are patterns of the cosinusoidal distribution $E_x = \cos(\pi y/b)$ with a far-field pattern as given in Table 3.1. The derivatives of $F(k_1)$ were obtained from tables (Milne, 1952) with a maximum of five terms in eqn. 4.16 used. Corresponding experimental results have also been reported (Jull, 1962, 1963). The patterns show that with a directive measuring antenna near-field sidelobe levels either above or below the far-field values may be observed. They also show the tendency for a highly directive antenna to record the far-field pattern in the near field. A practical difficulty of measuring far-field patterns at reduced ranges in this way is the interaction which occurs between measured and measuring antennas and which is not accounted for here.

There is a considerable more recent literature on near-field pattern measurements including numerical methods for including the effect of probe directivity in planar and cylindrical surface scanning techniques. Much of this is referred to by Joy *et al.* (1978) and Paris *et al.* (1978). Probe directivity compensation has also been applied in spherical scanning techniques (Jensen, 1975, Larson, 1977).

4.5 Far-field patterns from near-field measurements

Since the far-field pattern of an antenna is the Fourier transform of the tangential field in the aperture, it can be calculated from the measured amplitude and phase of the field in the antenna aperture. It may be convenient to base this calculation on measurements in a plane parallel to the aperture surface, rather than the aperture itself, and if this surface is sufficiently far from the antenna the problem of interaction between the probe and the antenna is avoided. Also the reactive fields of the aperture, which do not contribute to the far field pattern, have less effect on the observations and a more accurate prediction is possible. It will, of course, be necessary to measure the fields over an area larger than the aperture, which increases with the distance from the aperture, and only the pattern in the forward hemisphere can be obtained. At millimeter wavelengths, probe positioning errors may seriously affect the measured phase. However, this planar scanning technique is now a well developed practical method useful for high gain antennas particularly those which cannot be rotated (see, for example, Joy *et al.*, 1978).

If the antenna can be rotated measurements over a cylindrical or spherical surface may be more convenient. The amplitude phase and polarisation of the field on the surface of a circular cylinder axial to the antennas rotation will provide the lateral and rear, as well as the forward, patterns (Leach and Paris, 1973). For highly directive antennas, a combination of planar and cylindrical scanning may be used. Measurements over a spherical surface surrounding the antenna can in principle provide the complete pattern most efficiently (Wacker, 1975). This is, however, a formidable task, possible at present only for antennas not highly directive, but developments in measurement technique, numerical methods and especially computer technology are contributing towards its more general realisation (Hansen, 1980).

The procedure described here illustrates the method in which the far-field pattern can be obtained from the coefficients of the modal expansion of the near field. The discussion is restricted to two-dimensional fields but the results can be directly applied to obtain the principal plane pattern of an antenna large in one dimension and small in the other. The cylindrical coordinates r,θ are centered on the antenna aperture of width a of Fig. 4.3 and the electric field is entirely x-directed. The reader unfamiliar with cylindrical mode expansions of fields may refer to Appendix A.3. Outside a circle of radius $r = b/2$, the field can be written

$$E_x(r,\theta) = \sum_{n=-\infty}^{\infty} a_n H_n^{(2)}(kr) \exp(jn\theta), \qquad r > b/2 \tag{4.20}$$

In the far field, $kr \gg 1$ and

$$H_n^{(2)}(kr) \simeq \sqrt{\frac{2}{\pi kr}} \exp\left[-jkr + j\left(\frac{2n+1}{4}\right)\pi\right] \tag{4.21}$$

so that

$$E_x(r,\theta) \simeq \sqrt{\frac{2}{\pi kr}} \exp(-jkr + j\pi/4) F(\theta) \tag{4.22}$$

where

$$F(\theta) = \sum_{n=-\infty}^{\infty} a_n \exp[jn(\theta + \pi/2)] \tag{4.23}$$

is the radiation pattern of the antenna.

Suppose the amplitude and phase of the pattern is measured at a range $r = d$. Then,

$$E_x(d,\theta) = \sum_{-\infty}^{\infty} a_n H_n^{(2)}(kd) \exp(jn\theta) \tag{4.24}$$

This Fourier series may be inverted to yield the coefficients

$$a_n = \frac{1}{2\pi H_n^{(2)}(kd)} \int_0^{2\pi} E_x(d,\theta) \exp(-jn\theta) \, d\theta \tag{4.25}$$

The number of coefficients required in eqn. 4.23 depends on the size of the aperture. Only supergain antennas have coefficients a_n significantly different from zero for $n > kb/2$. Hence complex amplitudes a_n for n running from $-kb/2$ to $kb/2$, or $kb + 1$ in all, are required. The same minimum number of equally spaced observation points about the aperture are also needed.

The measuring antenna will have some directivity. If its effects is taken into account the observed pattern at $r = d$ will be

$$D(d,\theta) = \sum_{\infty}^{\infty} c_n(d) \exp(jn\theta) \tag{4.26}$$

where $c_n(d)$ is the response of the measuring antenna to the nth radial mode at $r = d$, i.e. to

$$a_n H_n^{(2)}(kd) \exp(jn\theta) \tag{4.27}$$

If the pattern of the receiving antenna pattern is assumed to be of the form of eqn. 4.14, a good approximation to $c_n(d)$ (Brown and Jull, 1961) is

$$c_n(d) = a_n \exp(-K)H_n^{(2)}(kd + jK) \tag{4.28}$$

where K is given by eqn. 4.15. Then the observed pattern is

$$D(d,\theta) = \sum a_n \exp(-K)H_n^{(2)}(kd + jK)\exp(jn\theta) \tag{4.29}$$

where

$$a_n = \frac{1}{2\pi \exp(-K)H_n^{(2)}(kd + jK)} \int_0^{2\pi} D(d,\theta)\exp(-jn\theta)d\theta \tag{4.30}$$

The coefficients a_n for calculation of the far-field pattern of eqn. 4.23 are thus readily obtained from a Fourier analysis of the measured near-field pattern. An

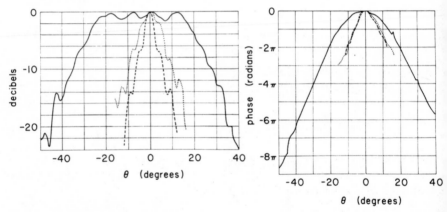

Fig. 4.5   *Amplitude and phase patterns of a 3·2 cm waveguide array measured with an essentially omnidirectional probe at ranges*
      ——— $r = 0.07 b^2 / \lambda$
      · · · · · $r = 0.17 b^2 / \lambda$
      — — — $r = 0.28 b^2 / \lambda$
      (After Martin, 1967) © 1967, IEEE

experimental demonstration of this technique for a 3·2 cm slotted waveguide array is shown in Figs. 4.5 and 4.6. Here Martin (1967) achieved better results from measurements very near the antenna ($r = 0.07 b^2 / \lambda$) than further out in the near field where the scan angle was smaller. These calculations assume no probe directivity, but an experimental and numerical demonstration in which this factor is included was given by Brown and Jull (1961).

The above method has been modified for application to three-dimensional cylindrical surfaces and applied to a paraboloidal reflector by Borgiotti (1978). His subsequent comments (Leach, Larson and Borgiotti, 1979) concerning the range of validity of the original development apply only to approximations of the Hankel function in eqn. 4.3. The method is valid at all ranges which encompass the measured antenna, provided interaction between probe and measured antenna is negligible, an effect not accounted for in any of the techniques proposed for determining the far-field pattern from near-field measurements.

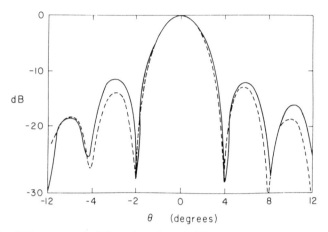

Fig. 4.6   *Far-field patterns of a 3·2 cm slotted waveguide array*
        —— measured
        − − − calculated from amplitude and phase measurements at a range $r = 0.07 b^2/\lambda$
        (see Fig. 4.5)
        (After Martin, 1967)

A thorough review of antenna pattern prediction from near-field measurements was made by Johnson *et al.* (1973), but there have been many contributions since, only a few of which have been referred to here.

Chapter 5

Aperture gain

5.1 Gain of aperture antennas

The directive gain of an antenna $G(\theta,\phi)$ is the ratio of its power density radiated in the θ,ϕ direction to the average of the power density radiated in all directions. If $\bar{E}, \bar{H}$ are the radiation fields of the antenna

$$G(\theta,\phi) = \frac{\frac{1}{2} \, \mathrm{Re}\{\bar{E}(r,\theta,\phi) \times \bar{H}^*(r,\theta,\phi) \cdot \hat{r}\}}{P_r/(4\pi r^2)} \tag{5.1}$$

where P_r is the total power radiated by the antenna, i.e.

$$P_r = \frac{1}{2} \, \mathrm{Re} \int_0^{2\pi} \int_0^\pi \bar{E}(r,\theta,\phi) \times \bar{H}^*(r,\theta,\phi) \cdot \hat{r} r^2 \, \sin^2\theta \, d\theta \, d\phi \tag{5.2}$$

For aperture antennas it is usually simpler to calculate this power by integrating over the aperture. Then for an aperture in the $z = 0$ plane

$$\begin{aligned} P_r &= \frac{1}{2} \, \mathrm{Re} \int_{Ap} \bar{E}(x,y) \times \bar{H}^*(x,y) \cdot \hat{z} \, dxdy \\ &= \frac{1}{2} \, \mathrm{Re} \int_{Ap} (E_x H_y^* - E_y H_x^*) \, dxdy \end{aligned} \tag{5.3}$$

If aperture field expressions of the form of eqn. 2.34 are inserted in eqn. 5.3, integration over homogeneous and inhomogeneous plane waves results. The integral over $k_x^2 + k_y^2 > k$ is related to the reactive energy stored in the evanescent waves and is excluded from this calculation by taking the real part of the integral. For apertures of sufficient size the radiating electric and magnetic fields are related by essentially free space conditions. Then eqn. 5.3 is

$$P_r = \frac{1}{2} \, Y_0 \int (|E_x(x,y)|^2 + |E_y(x,y)|^2) \, dxdy \tag{5.4}$$

In the far field of the aperture $\bar{H} = Y_0 \hat{r} \times \bar{E}$ and the time-averaged power flux is $\frac{1}{2} Y_0 |E(r,\theta,\phi)|^2$. For an in-phase aperture distribution the maximum directive gain is in the $\theta = 0$ direction. The IEEE standard definition of maximum directive gain is directivity, but it is more commonly referred to as simply gain. Then, using eqns. 2.48 and 5.4 in eqn. 5.1, the gain is

$$G = \frac{4\pi}{\lambda^2} \frac{|\int E_x(x,y)dxdy|^2 + |\int E_y(x,y)dxdy|^2}{\int |E_x(x,y)|^2 + |E_y(x,y)|^2 dxdy}$$

or

$$G = \frac{4\pi}{\lambda^2} \frac{|\int E_x(x,y)dxdy|^2}{\int |E_x(x,y)|^2 dxdy} \tag{5.5}$$

when the aperture distribution is polarised with $E_y = 0$. If it is uniform, $E_x(x,y)$ is a constant and

$$G = \frac{4\pi A}{\lambda^2} \tag{5.6}$$

where $A = ab$ for the rectangular aperture of Fig. 3.1, is the aperture area. In the Schwarz inequality

$$\left| \int fg\,dxdy \right|^2 \leqslant \int f^2 dxdy \int g^2 dxdy$$

let $f = E(x,y)$ and $g = 1$. Then

$$\left| \int E(x,y)dxdy \right|^2 \leqslant A \int \left| E(x,y) \right|^2 dxdy$$

Consequently, $G \leqslant 4\pi A/\lambda^2$ and for in-phase aperture distributions the uniform aperture field provides the highest gain.

5.2 Effective area

The transmitting and receiving radiation pattern, gain and impedance of an antenna free of nonreciprocal media are identical. This can be demonstrated from reciprocity (e.g. Collin and Zucker, 1969, p. 24, 98). The effective area of an antenna is defined as the ratio of the power available from the antenna to the incident power density for that component of the incident wave polarised in the direction of the radiation field of the antenna. The balance of the incident power is scattered from the antenna. Effective area and gain are related by reciprocity. Consider transmission between identically polarised matched antennas with gains G_1, G_2 and effective areas A_{e1}, A_{e2}. With power P_T fed to antenna 1, the power received by antenna 2 at a range r is

$$P_R = P_T G_1 \frac{1}{4\pi r^2} A_{e2} = P_T G_2 \frac{1}{4\pi r^2} A_{e1}, \tag{5.7}$$

where the right-hand-side of eqn. 5.7 is a consequence of reciprocity since P_R is also the power received by antenna 1 when P_T is fed to antenna 2. Hence $G_1 A_{e2} = G_2 A_{e1}$ and it is apparent that G/A_e is a constant for all antennas. The value of this constant is $4\pi/\lambda^2$. This comes from the result of eqn. 5.6 for an ideal aperture which receives all properly polarised power incident on it so that its effective area equals its physical area. Thus the coefficient of $4\pi/\lambda^2$ in eqn. 5.5 is the effective area and

$$A_e = \frac{\lambda^2 G}{4\pi} \tag{5.8}$$

For an aperture illuminated by a plane wave from the θ, ϕ direction, the effective area is $A_e(\theta, \phi) = \lambda^2 G(\theta, \phi)/4\pi$. The maximum effective area, eqn. 5.8, or simply effective area, as a fraction of the physical area is a convenient measure of aperture efficiency. It is about $0.4-0.6$ for horn antennas and for paraboloidal reflectors usually about 0.55, but as high as 0.75 for some specially shaped reflectors (Hansen, 1964).

5.3 Supergain antennas and aperture Q

If the concept of effective area is applied to a thin half-wave dipole, a value of $1.64\lambda^2/(4\pi)$ is obtained, much more than the physical area. For the short dipole, the physical area is greatly reduced but the collecting area is not less than that of a current element, $1.5\lambda^2/(4\pi)$. Reactive fields about the dipole are responsible for this apparent anomaly. Passing energy is guided in making the collecting area much larger than the physical area. As a consequence of these reactive fields, the impedance of the short dipole is largely reactive with a small radiation resistance. Unless this reactance is compensated for, the power gain of the antenna, which includes mismatch losses, remains substantially less than the directive gain. The dipole is an example of a supergain antenna, a term usually reserved for aperture antennas or arrays.

If the uniform aperture phase restriction is removed there is no theoretical limit to the gain of an aperture antenna. Such severe bandwidth limitations, tolerances and matching problems are usually encountered, however, that supergain arrays or aperture antennas are rarely used.

A supergain ratio for an aperture can be defined by relating its radiative and reactive power to its radiation pattern function. This relationship was derived by Borgiotti (1963, 1967), Collin and Rothschild (1963) and Rhodes (1966). The power radiated from an aperture in the $z = 0$ plane can be calculated from eqn. 5.2. With aperture fields $\overline{E}, \overline{H}$ given by eqns. 2.34 and 2.37, this is

$$P_r = \frac{Y_0}{8\pi^2 k} \iint\limits_{k_x^2 + k_y^2 < k^2} |F(k_x, k_y)|^2 \frac{(k^2 - k_y^2)\, dk_x dk_y}{\sqrt{k - k_x^2 - k_y^2}} \tag{5.9}$$

Similarly, the reactive power of the aperture obtained from the imaginary part of the complex Poynting vector integrated over the aperture is

$$P_j = \frac{Y_0}{8\pi^2 k} \int\int_{k_x^2 + k_y^2 > k^2} |F(k_x, k_y)|^2 \frac{(k_y^2 - k^2) dk_x dk_y}{\sqrt{k_x^2 + k_y^2 - k^2}} \tag{5.10}$$

Expressions for arbitrarily polarised aperture fields are given by Rhodes (1974, p. 47).[†]

For a resonant circuit a quality factor Q is defined as the ratio of the capacitive or inductive reactance of the circuit to its resistance and is a useful indicator of the frequency bandwidth of the circuit, the reciprocal of Q. A difficulty arises in applying this concept to antennas however, in that the stored electric and magnetic energies are individually singular. If these capacitive and inductive energies are taken together their singularities cancel. Thus aperture Q might be defined as the ratio of reactive to radiative power, i.e. P_j/P_r. This is unsatisfactory, however, for the net reactive power, and hence Q, could be small while the stored energies in the electric and magnetic fields are large. This situation led Rhodes (1974) to define aperture Q as the ratio of the net 'observable' reactive power to the radiated power. This 'observable' reactive power is

$$P_j' = \omega(W_E' + W_M') = \left\{ \frac{1}{\omega} \frac{\partial}{\partial \omega} [\omega(W_M' - W_E')] \right\}_{\omega_{res}} \tag{5.11}$$

where ω_{res} is the angular frequency at resonance and W_E' and W_M' are twice the nonsingular terms in the time-averaged stored electric and magnetic energy. Thus

$$W_M' - W_E' = \frac{1}{4} \int_V (\mu_0 \bar{H} \cdot \bar{H}^* - \epsilon_0 \bar{E} \cdot \bar{E}^*) dV \tag{5.12}$$

where the volume integration is over all space in $z > 0$. Then the antenna bandwidth is

$$B \approx \frac{1}{Q} = \frac{P_r}{P_j'} \tag{5.13}$$

and the supergain ratio γ is defined as

$$\gamma = 1 + Q = 1 + \frac{P_j'}{P_r} \quad . \tag{5.14}$$

For the fields of Section 2.6, P_j' is given by eqn. 5.10 with $(k_y^2 + k^2)$ replacing $k_y^2 - k^2$ in the numerator of the integrand.

The constraint implicit in eqn. 5.11, i.e. the frequency derivative must be evaluated at resonance, limits the applicability of Rhodes' definition. It applies well

---

[†] As the radiative and reactive power are directly proportional to the radiation resistance and the reactance, respectively, of the antenna, eqns. 5.9 and 5.10 show the impedance of an antenna can be derived from its radiation pattern function. This Poynting vector method of calculating radiation resistance is well known but its extension to reactance is relatively novel. It requires a pattern function that can be continued analytically from its visible region of real angles of propagation to its invisible region of complex angles. Rhodes (1964, 1974) has shown how the reactance of a planar dipole can be derived in this way.

to the planar dipole but in general it will be simpler and more satisfactory to obtain the frequency bandwidth of an antenna directly from the behaviour of its input reactance with frequency rather than from its Q (Collin, 1967).

5.4 Near-field and far-field axial gain

Expressions for the near-field and far-field axial gain of a few simple in-phase aperture distributions are developed in this section. For all but the smallest apertures electric and magnetic fields are related by $\bar{H} = Y_0 \hat{r} \times \bar{E}$ at ranges $\bar{r}$ well within the near-field or Fresnel zone of the antenna. Then the on-axis ($\theta = 0$) gain is, from eqn. 5.1,

$$G_n = \frac{\frac{1}{2} Y_0 |E_n(r,0)|^2}{P_r/(4\pi r^2)} \tag{5.15}$$

where $\bar{E}_n(r,0)$ is the axial field intensity and the subscript n denotes near field.

5.4.1 Uniform circular aperture
For a circular aperture of radius $a/2$ with a uniform distribution of unit amplitude and polarised in the x-direction, the on-axis near field is, from eqn. 3.27, with a quadratic phase variation in the aperture field to account for the finite range r,

$$\bar{E}_n(r,0) = \frac{\hat{\theta}j \exp(-jkr)}{\lambda r} \int_0^{2\pi} \int_0^{a/2} \exp(-j\beta\rho^2)\rho' d\rho' d\phi'$$

$$= \hat{\theta}2j \exp[-j(kr + t)] \sin t \tag{5.16}$$

where $\beta = \pi/r\lambda$ and $t = ka^2/(16r)$.
The axial power density is

$$\frac{1}{2}Y_0 |E_n(r,0)|^2 = 2Y_0 \sin^2 t \tag{5.17}$$

and varies with range as shown in Fig. 5.1a. As r decreases, the first maximum occurs at $r = a^2/(4\lambda)$ and the first minimum at $r = a^2/(8\lambda)$. At these axial ranges the circular aperture contains exactly one and two Fresnel zones, respectively, (Silver, 1949, p. 198) and there is, respectively, in-phase addition and complete cancellation of the axial field. The accuracy of this Fresnel zone approximation deteriorates seriously at axial ranges less than the first minimum (Hansen and Bailin, 1959).

The power radiated by an aperture of area A with a uniform polarised electric field of unit amplitude is $P_r = \frac{1}{2} Y_0 A$ watts. For this circular aperture, $A = \pi a^2/4$ and using this P_r and eqn. 5.17 in eqn. 5.15 yields the axial gain

$$G_n = \frac{4\pi A}{\lambda^2} \left(\frac{\sin t}{t}\right)^2 \tag{5.18}$$

In the far field, $r \to \infty$, $t \to 0$ and $G_n \to G$ given by eqn. 5.6. Thus $(\sin^2 t)/t^2$ is a near-field gain reduction factor for a circular aperture.

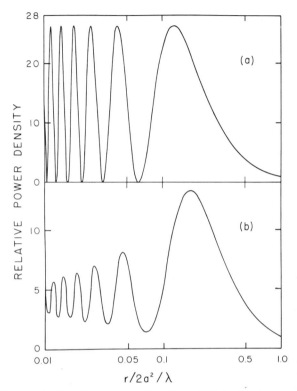

Fig. 5.1 *Axial power density of apertures with uniform field distributions*
a Circular aperture of diameter *a*
b Square aperture of side *a*

5.4.2 Uniform rectangular aperture

The field expressions of Section 4.2 for a rectangular aperture with uniform and cosinusoidal distributions simplify considerably on the beam axis where $k_1 = k_2 = 0$. In eqn. 4.7, $u_2 = -u_1 = a/\sqrt{2r\lambda}$ and, since $C(-u) = -C(u)$ and $S(-u) = -S(u)$, eqn. 4.6 becomes

$$D_1(0) = \sqrt{2r\lambda} \, [C(u) - jS(u)] \tag{5.19}$$

For a uniformly illuminated rectangular aperture $D_2(0)$ is given by eqn. 5.19 with $v = b/\sqrt{2r\lambda}$ replacing u. The on-axis near field of the aperture of Fig. 3.1 is, from eqns. 4.4 and 5.19,

$$\bar{E}_n(r,0) = \hat{\theta} 2j \exp(-jkr)[C(u) - jS(u)] \, [(C(v) - jS(v)] \tag{5.20}$$

As the power radiated is $P_r = \frac{1}{2} Y_0 A$, where $A = ab$, the axial gain can be written as

$$G_n = \frac{4\pi A}{\lambda^2} R_E(u) R_E(v) \tag{5.21}$$

where

$$R_E(u) = \frac{C^2(u) + S^2(u)}{u^2} \tag{5.22}$$

with $u = a/\sqrt{2r\lambda}$ and $C(u)$ and $S(u)$ defined by eqn. 4.8. The gain reduction factor $R_E(u)$ for a uniform line source was first tabulated by Polk (1956). Its behaviour is illustrated in Fig. 5.2 and values are given in Table 5.1, as functions of the parameter $2d^2/(\lambda r)$ where d is the aperture dimension a and $\Re$ is the range r. As $r \to \infty$, $u \to 0$, $C(u) \approx u$, $S(u) \approx u^3 \approx 0$ and $R_E(u) \to 1$; then eqn. 5.21 reduces to eqn 5.6.

The near-field axial gain of a square aperture of side a with a uniform distribution is shown in Fig. 5.1b where its axial power density

$$2Y_0 [C^2(u) + S^2(u)]^2 \tag{5.23}$$

is plotted with the power density normalised to unity at $r = 2a^2/\lambda$. Incomplete cancellation of axial power density and gain occurs at ranges r for which, with a circular aperture, there is an even integral multiple of Fresnel zones in the aperture.

5.4.3 Uniform and cosinusoidal rectangular aperture

If the electric field is $E_x(x,y) = \cos(\pi y/b)$ in the aperture of Fig. 3.1, eqn. 4.9 with $k_2 = 0$ is used for the axial field. Then, from eqn. 4.10,

$$\left. \begin{array}{l} w_1 = -v_2 = w = \dfrac{-b}{\sqrt{2r\lambda}} + \dfrac{1}{b}\sqrt{\dfrac{r\lambda}{2}} \\[2ex] w_2 = -v_1 = v = \dfrac{b}{\sqrt{2r\lambda}} + \dfrac{1}{b}\sqrt{\dfrac{r\lambda}{2}} \end{array} \right\} \tag{5.24}$$

and

$$D_2(0) = \sqrt{\frac{r\lambda}{2}} \exp[j(\pi r\lambda/4b^2)]\{C(v) - C(w) - j[S(v) - S(w)]\} \tag{5.25}$$

The on-axis near field of a rectangular aperture with uniform and cosinusoidal field distributions is

$$\bar{E}_n(r,0) = \hat{\theta} j \exp[-jkr + j(\pi r\lambda/4b^2)]$$
$$\times [C(u) - jS(u)]\{C(v) - C(w) - j[S(v) - S(w)]\} \tag{5.26}$$

which with the power radiated

$$P_r = \frac{1}{2} Y_0 \int_{-a/2}^{a/2} \int_{-b/2}^{b/2} \cos^2\left(\frac{\pi y}{b}\right) dy\,dx = \frac{1}{4} Y_0 ab, \tag{5.27}$$

in eqn. 5.15 gives the axial gain

$$G_n = \frac{32A}{\pi\lambda^2} R_E(u) R_H(v,w) \tag{5.28}$$

In eqn. 5.28 the aperture area $A = ab$, $R_E(u)$ is defined by eqn 5.22 and

$$R_H(v,w) = \frac{\pi^2}{4} \frac{[C(v) - C(w)]^2 + [S(v) - S(w)]^2}{(v - w)^2} \tag{5.29}$$

is the near-field gain reduction factor for a linear cosinusoidal distribution. The behaviour of $R_H(v,w)$ with range r is illustrated in Fig. 5.2 and its values tabulated in Table 5.1 where in the parameter $s = 2d^2/\lambda l$, $d = b$ or a and $l = r$. As $r \to \infty$, both $R_e(u)$ and $R_H(v,w)$ become unity and eqn. 5.28 is

$$G = \frac{32A}{\pi\lambda^2} \tag{5.30}$$

the axial gain of a rectangular aperture with uniform and cosinusoidal distributions.

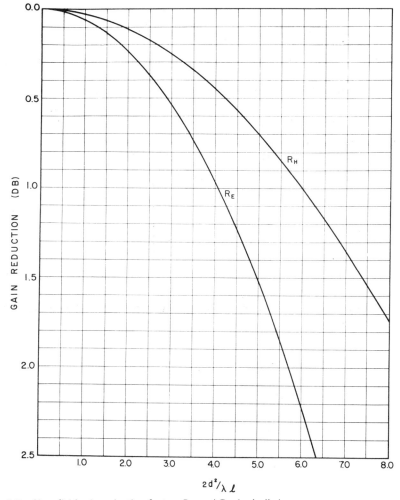

Fig. 5.2 *Near-field gain reduction factors R_E and R_H in decibels*

Table 5.1 *Gain reduction factors R_E and R_H in decibels. Interpolation using second differences will be in error by much less than 0·01 dB*

$2d^2/\lambda$	R_E	R_H
0·5	0·015	0·007
1·0	0·060	0·029
1·5	0·134	0·064
2·0	0·239	0·114
2·5	0·374	0·179
3·0	0·541	0·257
3·5	0·738	0·349
4·0	0·967	0·454
4·5	1·229	0·573
5·0	1·525	0·705
5·5	1·854	0·850
6·0	2·218	1·007
6·5	2·618	1·176
7·0	3·054	1·357
7·5	3·527	1·547
8·0	4·037	1·748

5.4.4 Axial gain reduction

A comparison between near-field gain reductions for circular and square uniformly illuminated apertures and a square aperture with uniform and cosinusoidal illuminations is shown in Table 5.2. These values, obtained from eqns. 5.18, 5.21 and

Table 5.2 *Effective areas and axial gain reductions of uniform in-phase aperture distributions*

A_e	$\dfrac{\pi a^2}{4}$	a^2	$\dfrac{8a^2}{\pi^2}$
$r = a^2/\lambda$	0·950	0·896	0·922
$r = 2a^2/\lambda$	0·987	0·973	0·980

5.28 are for a circular aperture of diameter a and square apertures of side a. Near-field gain reduction depends on the largest aperture dimension as well as the aperture distribution. Here the aperture distribution with the highest gain or effective area exhibits the largest near-field gain reduction. This will generally hold for in-phase distributions symmetrically tapered from the aperture center to the edges.

The results of this Section are for point measuring probes in the near-field. If the measuring probe directivity is significant axial near-field gain can be specified only for identical antennas. Near-field gain reduction factors for identical circular and square apertures with uniform and tapered distributions have been derived (e.g. Soejima, 1963; Hansen, 1964).

Chapter 6

Applications of aperture theory to antennas

6.1 Approximations and limitations

A few examples of the application of Fourier transform theory to the calculation of the radiation properties of some practical antennas are given here. These examples are chosen to illustrate the method and nature of the approximations necessary in its application. The major approximation of the Kirchhoff or Fourier transform method is the assumed form of the field in the aperture plane both inside and outside the aperture. Usually the tangential electric field is assumed zero outside the aperture and the incident field in the aperture. With eqn. 2.42 or 3.1 it is also assumed that tangential electric and magnetic fields in the aperture are related by plane wave conditions. All these approximations tend to improve in accuracy as the aperture dimensions increase in wavelengths. It is understood, however, that only radiation in the forward hemisphere can be calculated by the method and that if the boundary conditions in the aperture plane are not satisfied errors in the calculated field increase near the boundary. Generally the accuracy of the result decreases as the angle off the beam axis increases. For large apertures, however, calculation of the main beam and first few sidelobes of the pattern does not involve large angles off the beam axis and in this range high accuracy can be expected.

6.2 Open-ended rectangular waveguides

The simplest practical aperture antenna, the open-ended rectangular waveguide supporting only the dominant mode is, unfortunately, difficult to analyse accurately. If the Kirchhoff method is used then for the TE_{01} waveguide mode the incident field in the aperture of Fig. 6.1 is

$$E_x(x,y) = E_0 \cos\left(\frac{\pi y}{b}\right) \tag{6.1}$$

and the principal plane radiation patterns obtained from eqn. 3.3 are given by eqn. 3.6 and the pattern of the cosinusoidal distribution of Table 3.1. This result is useful for large open-ended waveguides but is a crude approximation for waveguides supporting only the TE_{01} mode, for then $\lambda/2 < b < \lambda$ and $a \approx b/2$, dimensions usually considered too small for a valid application of this method.

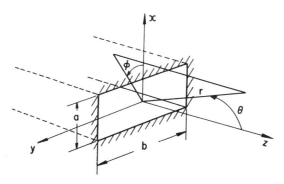

Fig. 6.1   *Rectangular waveguide aperture in a conducting plane*

For a waveguide aperture in a conducting screen occupying the remainder of the $z = 0$ plane of Fig. 6.1 the accuracy of the method is much better, for the boundary condition $E_x = 0$ is rigorously satisfied outside the aperture. If eqn. 2.38 is used to calculate the radiation field the only approximation is the assumption of eqn. 6.1 as the aperture field. This is in error near the edges $x = \pm a/2$, $z = 0$, where the electric field is singular, but the boundary condition at the other pair of edges is satisfied.

Lewin (1976) has shown how the effect of the conducting screen can be considered as a modification of the Huygens' obliquity factor $(1 + \cos \theta)$ in eqn. 2.42. This will lead to the same result as that obtained here. With eqn. 6.1 in eqn. 2.38, the radiation field of the rectangular aperture in $z > 0$ is

$$\bar{E}(r,\theta,\phi) = \frac{j \exp(-jkr)}{\lambda r} (\hat{\theta} \cos \phi - \hat{\phi} \sin \phi \cos \theta) E_0 a \frac{\sin(k_1 a/2)}{k_1 a/2}$$

$$\times 2\pi b \frac{\cos(k_2 b/2)}{\pi^2 - (k_2 b)^2} \tag{6.2}$$

where k_1 and k_2 are defined by eqn. 3.5.

Following Collin and Zucker (1969, p. 562), it is instructive to examine this radiation field for a narrow half-wave slot in a conducting screen. With $ka \ll 1$ and $kb = \pi$, eqn. 6.2 becomes

$$\bar{E}(r,\theta,\phi) = \frac{jV_0}{\pi r} \exp(-jkr)(\hat{\theta} \cos \phi - \hat{\phi} \sin \phi \cos \theta) \frac{\cos \{(\pi/2) \sin \theta \sin \phi\}}{1 - \sin^2 \theta \sin^2 \phi}$$

$$\tag{6.3}$$

where $V_0 = E_0 a$ is the voltage across the slot. In the $\phi = 0$ plane, $E_\phi = 0$ and

$$E_\theta = \frac{jV_0}{\pi r} \exp(-jkr) \tag{6.4}$$

an omnidirectional pattern. In the $\phi = \pi/2$ plane, $E_\theta = 0$ and

$$E_\phi = \frac{-jV_0 \exp(-jkr)}{\pi r} \frac{\cos\{(\pi/2)\sin\theta\}}{\cos\theta} \tag{6.5}$$

These expressions have the same form as the field of a complementary $\lambda/2$ dipole along the y-axis, with a change in polarisation. The magnetic field of the dipole corresponds to the electric field of the slot in accordance with the electromagnetic form of Babinet's principle (Booker 1946).

That the gain and impedance of complementary dipole and slot are also related can be shown by calculating the total power radiated by the slot.

$$P_r = \frac{1}{2}Y_0 \int_0^{2\pi} \int_0^{\pi/2} (|E_\theta|^2 + |E_\phi|^2)r^2 \sin\theta \, d\theta d\phi$$

$$= \frac{Y_0}{2}\left(\frac{V_0}{\pi}\right)^2 \int_0^{2\pi} \int_0^{\pi/2} \frac{\cos^2\{(\pi/2)\sin\theta\sin\phi\}}{1 - \sin^2\theta\sin^2\phi} \sin\theta \, d\theta d\phi \tag{6.6}$$

where eqn. 6.3 has been used. In the coordinates r, ψ, γ, with the y-axis of Fig. 6.1 as the polar axis, $\sin\theta\sin\phi = \cos\psi$ and the integral is

$$P_r = \frac{Y_0}{2}\left(\frac{V_0}{\pi}\right)^2 \int_{-\pi/2}^{\pi/2} \int_0^{\pi} \frac{\cos^2\{(\pi/2)\cos\psi\}}{\sin\psi} d\psi d\gamma \tag{6.7}$$

which reduces to (e.g. Jordan and Balmain, 1968, p. 330)

$$P_r = \frac{Y_0 V_0^2}{4\pi} \int_0^{2\pi} \frac{1 - \cos v}{v} dv = Y_0 \frac{V_0^2}{4\pi} \text{Cin}(2\pi) \tag{6.8}$$

The axial ($\theta = \phi = 0$) gain is

$$G = \frac{\frac{1}{2}Y_0|E_0(0)|^2}{P_r/(4\pi r^2)} = \frac{8}{\text{Cin}(2\pi)} = 3 \cdot 28 \tag{6.9}$$

or $5 \cdot 16$ dB. (For a table of the Cin function see Abramowitz and Stegun, 1964, p. 244.) This is twice the gain of a thin half-wave dipole, appropriately, since the slot radiates only into the half-space $z > 0$.

The radiation conductance of this slot is

$$G_r = \frac{2P_r}{V_0^2} = \frac{\text{Cin}(2\pi)}{240\pi^2} = 1 \cdot 03 \times 10^{-3} \text{ S} \tag{6.10}$$

or half the value obtained from Booker's (1946) expression for the admittance $Y_s = 4Y_0^2 Z_d$ of a slot radiating on both sides of the conducting plane with $Z_d = 73 \cdot 1 \ \Omega$ as the impedance of a thin half-wave dipole.

6.3 Pyramidal horns

6.3.1 Aperture field

The most commonly used microwave antenna, the pyramidal horn cannot be analysed rigorously. As the adjoining sides of the horn do not constitute a pair of orthogonal surfaces, it is not possible to express the fields in the horn exactly as a sum of outgoing and incoming spherical modes, as can be done for a conical horn. Even in the latter case, however, determination of the mode amplitudes resulting from diffraction at the aperture is difficult. The following treatment, which neglects diffraction across the aperture and aperture-throat interaction, effects considered in Chapter 9, yields satisfactory results for most purposes, however.

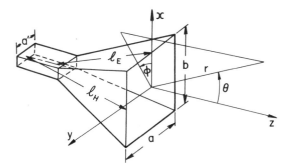

Fig. 6.2   *Coordinates of a pyramidal horn*

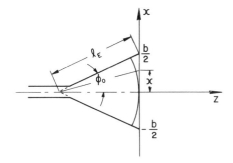

Fig. 6.3   *E-plane path differences in a horn*

It is assumed that in the waveguide feed to the horn of Fig. 6.2 only the dominant (TE_{01}) mode is present. The path length from the horn apex to the aperture plane in the E-plane, or the plane containing the electric field vector, is indicated in Fig. 6.3. This path length change with aperture position results in a phase variation in the aperture field, which in the E-plane of the horn is

$$k[(l_E \cos \phi_0)^2 + x^2]^{1/2}$$

$$\approx kl_E \cos \phi_0 + \frac{kx^2}{2l_E \cos \phi_0} , \quad x^2 \ll l_E \cos \phi_0 \tag{6.11}$$

or approximately $kx^2/(2l_E)$ with phase referred to the aperture centre and ϕ_0 small. Similarly, the H-plane phase variation of the aperture field is $ky^2/(2l_H)$. Hence the aperture field distribution is approximately,

$$E_x(x,y) = E_0 \cos\left(\frac{\pi y}{a}\right) \exp\left[-jk\left(\frac{x^2}{2l_E} + \frac{y^2}{2l_H}\right)\right] \tag{6.12}$$

where E_0 is the field at the aperture centre.

6.3.2 Radiation patterns

When eqn. 6.12 is used in eqn. 3.1, the far-field radiation pattern is

$$\overline{E}(r,\theta,\phi) = \overline{A}E_0 I_1(k_1)I_2(k_2) \tag{6.13}$$

where

$$I_1(k_1) = \int_{-b/2}^{b/2} \exp[-j(\beta_1 x^2 - k_1 x)] \, dx \tag{6.14}$$

$$I_2(k_2) = \int_{-a/2}^{a/2} \cos\frac{\pi y}{b} \, \exp[-j(\beta_2 y^2 - k_2 y)] \, dy \tag{6.15}$$

with $\beta_1 = \pi/(\lambda l_E)$, $\beta_2 = \pi/(\lambda l_H)$ and k_1 and k_2 defined by eqn. 3.5. Reduction of these integrals to tabulated Fresnel integrals is given by eqn. 4.6 for 6.14 and by eqn. 4.9 for 6.15 with l_E and b replacing r and a, respectively, in eqns. 4.6 and 4.7, and l_H and a replacing r and b in eqns. 4.9 and 4.10.

The normalised E-plane ($\phi = 0$) radiation pattern is

$$\frac{E_\theta(r,\theta)}{E_\theta(r,0)} = \frac{(1 + \cos\theta)}{2} \frac{I_1(k\sin\theta)}{I_1(0)}$$

$$= \exp\left(j\frac{\pi l_E}{\lambda}\sin^2\theta\right) \frac{(1 + \cos\theta)}{4}$$

$$\times \frac{\{C(u_2) - C(u_1) - j[S(u_2) - S(u_1)]\}}{C(u) - jS(u)}, \tag{6.16}$$

where

$$u = \frac{b}{\sqrt{2\lambda l_E}} \tag{6.17}$$

$$u_2 \atop 1 = \pm u - \sqrt{\frac{2l_E}{\lambda}} \sin\theta \tag{6.18}$$

and the Fresnel integrals are defined by eqn. 4.8. In deriving eqn. 6.16, $C(-u) = -C(u)$ and $S(-u) = -S(u)$ were used.

From eqn. 6.16, the relative power pattern is

$$\left| \frac{E_\theta(r,\theta)}{E_\theta(r,0)} \right|^2 = \left(\frac{1 + \cos\theta}{4} \right)^2 \frac{[C(u_2) - C(u_1)]^2 + [S(u_2) - S(u_1)]^2}{C^2(u) + S^2(u)}$$

$$= \left(\frac{1 + \cos\theta}{\pi} \right)^2 \frac{R_H(u_2, u_1)}{R_E(u)} \qquad (6.19)$$

where R_E and R_H are defined by eqns. 5.22 and 5.29.

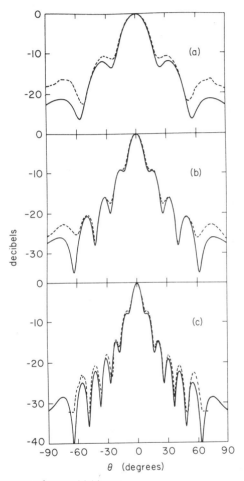

Fig. 6.4   *E-plane patterns of pyramidal horns*
——— calculated from eqn. 6.19
- - - - measured (Slayton, 1954)
a $b = 2 \cdot 40\lambda$, $1_E = 4 \cdot 21\lambda$
b $b = 4 \cdot 50\lambda$, $1_E = 10 \cdot 00\lambda$
c $b = 6 \cdot 67\lambda$, $1_E = 18 \cdot 52\lambda$

E-plane radiation patterns of pyramidal horns calculated from eqn. 6.19 are shown in Fig. 6.4. Also shown for comparison are measured patterns of these horns, widely used as microwave antenna gain standards (Slayton, 1954). As expected, eqn. 6.19 is most accurate for small angles θ and the main beam is well predicted in each of these examples. This accuracy improves with increasing aperture size, as is evident at angles well off the beam axis.

The H-plane ($\phi = \pi/2$) pattern of a pyramidal horn is

$$\frac{E_\phi(r,\theta)}{E_\phi(r,0)} = \frac{(1 + \cos \theta)}{2} \frac{I_2(k \sin \theta)}{I_2(0)} \tag{6.20}$$

where $I_2(k \sin \theta)$ is given by eqn. 4.9 with a and l_H replacing b and r, respectively, in eqns. 4.9 and 4.10, and

$$I_2(0) = \sqrt{\frac{\lambda l_H}{2}} \, \exp\left(j \, \frac{\lambda l_H \pi}{4a^2}\right) \{C(v) - C(w) - j[S(v) - S(w)] \} \tag{6.21}$$

with

$$\frac{v}{w} = \pm \frac{a}{\sqrt{2\lambda l_H}} + \frac{1}{a} \sqrt{\frac{\lambda l_H}{2}} \tag{6.22}$$

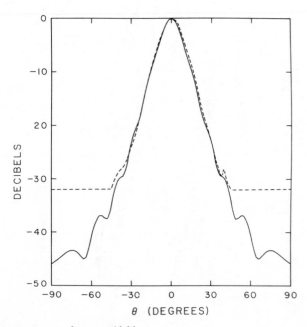

Fig. 6.5   *H-plane patterns of a pyramidal horn*
——— calculated from eqn. 6.20
- - - - measured (Slayton, 1954)
$a = 6.80\lambda$, $l_H = 10.70\lambda$

The H-plane patterns of the horns of Fig. 6.4 are essentially single lobed and are well predicted by eqn. 6.20, as indicated by the example of Fig. 6.5, but for smaller horns less accuracy can be expected of eqn. 6.20 than 6.16 as diffraction by the E-plane aperture edges can substantially affect the H-plane pattern.

6.3.3 Near-field patterns

As first-order effects of horn flare on the aperture distribution and finite range on the predicted pattern are similar, the horn patterns of Figs. 6.4a, b, and c are similar to the near-field patterns of uniform in-phase distributions in apertures of width b at ranges $0 \cdot 73\, b^2/\lambda$, $0 \cdot 49\, b^2/\lambda$ and $0 \cdot 42\, b^2/\lambda$, respectively. Near-field patterns of horns can be calculated by combining these effects. From Section 4.2, if the phase factor

$$\exp\left[-\frac{\pi}{r\lambda}(x^2 + y^2)\right] \tag{6.23}$$

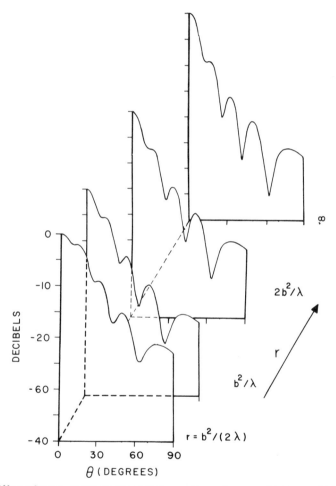

Fig. 6.6 *Effect of range on the predicted E-plane pattern of a pyramidal horn with* $b = 4 \cdot 50\lambda$, $1_E = 10 \cdot 00\lambda$

is included in the aperture field of eqn. 6.12, the near-field principal plane patterns of a pyramidal horn are given by eqns. 6.13–6.22 with l_E replaced by

$$l_E' = \frac{rl_E}{r + l_E} \tag{6.24}$$

and

$$l_H' = \frac{rl_H}{r + l_H} \tag{6.25}$$

replacing l_H. The effects of a finite range on the E-plane pattern of Fig. 6.4b are illustrated in Fig. 6.6. These effects are analogous to decreasing the flare length of the horn l_E with a fixed aperture width b. The main beam broadens, nulls are filled and sidelobe levels rise.

6.3.4 Pyramidal horn gain

As horn flare and finite range similarly affect the radiation patterns of in-phase aperture distributions the axial gain of a pyramidal horn has the same form as the axial near-field gain of in-phase uniform and cosinusoidal distributions. Thus, eqn. 5.28 also represents a convenient form of Schelkunoff's (1943) equation for the gain of a pyramidal horn.

$$G = G_0 R_E(u) R_H(v,w) \tag{6.26}$$

where $G_0 = 32ab/(\pi\lambda^2)$ is the far-field gain of an aperture with an in-phase uniform and cosinusoidal distribution (see eqn. 5.30) and $R_E(u)$, defined by eqn. 5.22 with $u = b/\sqrt{2\lambda l_E}$, is its reduction due to the E-plane flare of the horn. $R_H(v,w)$, defined by eqn. 5.29 with v and w given by eqn. 6.23, now represents the gain reduction due to the H-plane horn flare.

In Table 5.1, the values of R_E and R_H are tabulated in decibels as functions of the generalized parameter $s = 2d^2/(\lambda l)$, where d is the aperture dimension b or a and l is l_E or l_H, respectively. The range $0 < s < 8$ of Fig. 5.2 is sufficient for most horns. Interpolation from the Table 5.1 in this range is accurate to better than 0·01 db.

The gain of a pyramidal horn in decibels is

$$G_{dB} = 10\cdot084 + 10\log_{10}(ab/\lambda^2) + 10\log_{10}R_E + 10\log_{10}R_H \tag{6.27}$$

As an example, a 10 cm band standard pyramidal horn (Slayton, 1954) has dimensions $a = 3\cdot67\lambda$, $b = 2\cdot72\lambda$, $l_H = 5\cdot38\lambda$ and $l_E = 4\cdot78\lambda$ at a wavelength $\lambda = 8\cdot82$ cm. The gain of the in-phase aperture is $10\cdot08 + 10\log_{10}(3\cdot67 \times 2\cdot72) = 20\cdot08$ dB. For the E-plane flare, $2b^2/(\lambda l_E) = 3\cdot10$ and, from Table 5.1, $10\log_{10} R_E = -0\cdot58$ dB. For the H-plane flare, $2a^2/(\lambda l_H) = 5\cdot02$ and $10\log_{10} R_H = -0\cdot71$ dB. The far-field gain of the horn is thus $20\cdot08 - 0\cdot58 - 0\cdot71 = 18\cdot79$ dB. The measured value is $18\cdot87$ dB (Jull, 1968). Fig. 6.7 shows the predicted and measured variation in the gain of this horn with wavelength, the monotonic curve representing eqn. 6.26 or 6.27. The oscillating curve includes, approximately, diffraction and reflection within the horn, as described in Section 9.5.

An accuracy of about $\pm 0\cdot1$ dB for eqn. 6.27 is typical with well made horns and gains above about 17 dB. Few experimental values are available for less directive horns, but the accuracy of these formulas is worse, the predicted gain being less

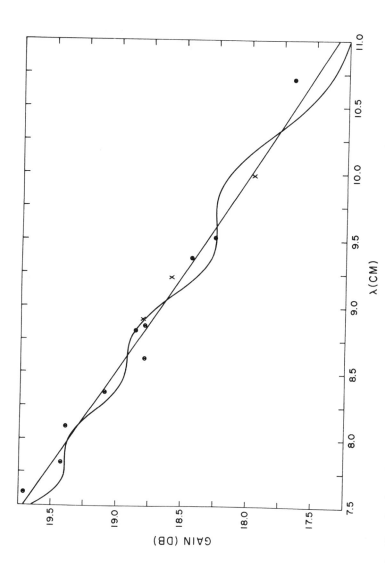

Fig. 6.7 *Gain of a 10 cm band standard pyramidal horn with dimensions a = 32·41 cm, b = 24·00 cm, l_E = 42·15 cm, l_H = 47·45 cm (see Fig. 6.2)*
Solid monotonic curve represents eqn. 6.26 or 6.27. Oscillating curve includes approximately diffraction and reflection in the *E*-plane of the horn (eqn. 9.96). Measured values: ⊗, X Slayton (1954) ● Jull and Deloli (1964). © 1973 IEEE (Jull, 1973b).

than the measured gain. For higher gain millimeter wavelength horns attenuation in the waveguide feed may amount to about 0·1 dB and unless this is accounted for the gain predicted with eqn. 6.27 will be higher than the measured gain.

The axial gain of a pyramidal horn at a finite range r may also be obtained from the above expressions by replacing l_E and l_H by eqns. 6.24 and 6.25, respectively (Jull, 1970). If the gain of the above horn is required at range $r = 2a^2/\lambda = 27·01\lambda$ for example, $l_H' = 4·49\lambda$ replaces l_H and $l_E' = 4·06\lambda$ replaces l_E. Then $2a^2/(\lambda l_H') = 6·02, 2b^2/(\lambda l_E') = 3·65$ and the combined flare and near-field gain reduction factors are $-1·01$ dB and $-0·81$ dB. The horn gain is now $20·08 - 1·82 = 18·26$ dB, or 0·53 dB less than its far-field value. This near-field reduction is substantially larger than the value of 0·09 dB predicted for an in-phase aperture at this range (see Table 5.2). It also exceeds half the reduction predicted for transmission between identical pyramidal horns at this range (Chu and Semplak, 1965).

6.4 Sectoral horns

6.4.1 Radiation patterns
Fig. 6.8a shows an E-plane sectoral horn; a horn flared only in the plane of its electric field vector. Its aperture field varies in phase only along the b-dimension

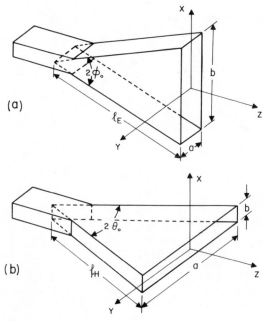

Fig. 6.8   *Coordinates of:*
   a E-plane sectoral horn
   b H-plane sectoral horn

and is given approximately by eqn. 6.12 with $l_H \to \infty$. The E-plane pattern is approximately given by eqn. 6.16 and the H-plane pattern is approximately that for an in-phase cosinusoidal distribution (Table 3.1 and Fig. 3.2). As the E-plane pattern of a horn is affected little by its H-plane dimensions, the patterns of Figs. 6.4 and 6.5 apply to both E-plane sectoral and to pyramidal horns. The pattern of a cosinusoidal distribution is a crude approximation to the H-plane pattern of most E-plane sectoral horns, which typically have narrow aperture dimensions $\lambda/2 < a < \lambda$.

The H-plane sectoral horn of 6.8b is a horn flared only in the plane of its magnetic field vector. Its E-plane pattern is approximated by the pattern of a uniform in-phase distribution of width b, i.e. by eqn. 3.8. The accuracy of the approximation of eqn. 6.20 for its H-plane pattern is impaired by diffraction from the E-plane edges.

As stated in Section 6.1, these expressions assume aperture electric and magnetic fields related by free-space plane wave conditions; i.e. by eqn. 2.15, an approximation which improves with aperture size. This assumption is not essential to the application of the Kirchhoff method and is not used in the vector diffraction integrals for the pattern given by Silver (1949, pp. 357–358) and corrected for sign errors in the E-plane horn expressions by Narasimhan and Rao (1973). Their examples are for horns shorter and with wider flare angles than those to which the expressions of Section 6.3.1 apply.

6.4.2 Sectoral horn gain
An expression for the axial gain of an E-plane sectoral horn, also due to Schelkunoff (1943), is obtained by letting $l_H \to \infty$ in eqn. 6.23. Then $R_H \to 1$ in eqn. 6.26 and the gain is

$$G = G_0 R_E(u) \tag{6.28}$$

The equivalent of this simple formula has appeared in many textbooks over the past four decades with no indication of its accuracy. It is inaccurate for most practical horns. For example, an E-plane sectoral horn with the same $b = 2 \cdot 72\lambda$ and $l_E = 4 \cdot 78\lambda$ dimensions at $\lambda = 8 \cdot 82$ cm as the pyramidal horn example and $a = 0 \cdot 818\lambda$, corresponding to the width of its waveguide feed, is not a particularly small horn. However its gain, predicted from eqn. 6.27 with the final term zero, is $10 \cdot 084 + 10 \log_{10} (0 \cdot 818 \times 2 \cdot 72) - 0 \cdot 580 = 12 \cdot 98$ dB, or $0 \cdot 85$ dB, below an accurately measured value of $13 \cdot 83$ dB. (Jull and Allan, 1974). Fig. 6.9 shows that as the frequency changes the measured gain of this horn is consistently higher than the value predicted by eqn. 6.28 (solid curve). A still larger discrepancy may be expected for other E-plane sectoral horns.

Even with vector Kirchhoff theory and virtually all other approximations removed the gain predicted, indicated by the dashed curve in Fig. 6.9, is unsatisfactory. Kirchhoff theory fails here because the aperture dimension a is too small. As the phase propagation constant in the horn is inaccurately approximated by its free space value $2\pi/\lambda$ the aperture field phase variation is incorrect; the TE_{10} mode propagation $2\pi/\lambda_g$ should be used instead. Also interaction across the a-dimension of the aperture needs to be accounted for.

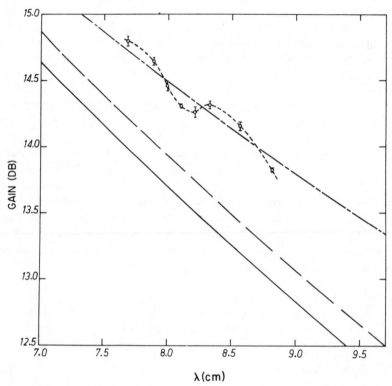

λ(cm)

Fig. 6.9   *Gain of an E-plane sectoral horn with* $a = 7.21$ *cm,* $b = 24.00$ *cm,* $l_E = 42.15$ *cm,* $2\phi_0 = 33°$ *and fed by 17.8 cm of WR(284) waveguide and a 25.4 cm waveguide tuner.*
——— eqn. 6.28
—·—· eqn. 6.34
--ǫ--ǫ-- measured values © 1974 IEEE (Jull & Allan, 1974).

This interaction may be included by noting that for rectangular apertures with separable E- and H-polarised fields the axial gain can be written

$$G = \frac{1}{\pi} G_E G_H \tag{6.29}$$

where G_E and G_H are two-dimensional axial gain solutions for apertures with dimensions in the plane of the electric and magnetic field vectors, respectively. Here G_H is the axial gain of an open-ended parallel-plate waveguide of width a supporting the TE_1 mode. From eqn. 9.50, this is exactly

$$G_H = \frac{8a}{\lambda(1 + \lambda_g/\lambda)} \exp\left[\frac{\pi a}{\lambda}\left(1 - \frac{\lambda}{\lambda_g}\right)\right]$$ (6.30)

where

$$\lambda_g = \frac{\lambda}{\sqrt{1 - (\lambda/2a)^2}}$$ (6.31)

is the guide wavelength.

G_E, the axial gain of a two-dimensional E-plane sectoral horn of aperture b and flare length l_E, is

$$G_E = \frac{2\pi}{\lambda} R_E(u)$$ (6.32)

where $R_E(u)$ is defined by eqn. 5.22. For a TEM horn u is given by eqn. 6.17, but here we should use instead

$$u = \frac{b}{\sqrt{2\lambda_g l_E}}$$ (6.33)

for the correct gain reduction due to the E-plane flare. Eqns. 6.32 and 6.30 in eqn. 6.29 give, for the gain of an E-plane sectoral horn (Jull and Allan, 1974),

$$G = \frac{16ab}{\lambda^2(1 + \lambda_g/\lambda)} R_E(u) \exp\left[\frac{\pi a}{\lambda}\left(1 - \frac{\lambda}{\lambda_g}\right)\right]$$ (6.34)

For the example considered previously, this yields 13·93 dB or only 0·1 dB below the measured value. The broken curve in Fig. 6.9, which passes through the median of the measured values, represents eqn. 6.34. The oscillation of about ±0·1 dB amplitude of the measurements in both Figs. 6.7 and 6.9 about the predicted values is due to the reflection of fields diffracted by the aperture from the interior surfaces of the horn and is difficult to account for quantitatively (Jull, 1973b).

Schelkunoff's (1943) expression for the axial gain of an H-plane sectoral horn is obtained from eqn. 6.26 by letting $l_E \to \infty$ in the arguments u of eqn. 6.17. Then $R_E(u) \to 1$ and the gain is

$$G_H = G_0 R_H(v,w)$$ (6.35)

where G_0, R_H, v and w are defined as before. Eqn. 6.35 is also widely quoted in other forms. Although there are reasons to believe it is more accurate than eqn. 6.28, evidently it has not been verified experimentally and so should be used with appropriate caution.

The near-field gain of sectoral horns can be calculated from eqns. 6.34 and 6.35 by the substitutions of eqns. 6.24 and 6.25 for l_E and l_H in eqns. 6.33 and 6.22.

Ohmic or current losses are usually negligible in horns. Mismatch losses reduce the gain by a factor $1 - |\Gamma|^2$, where Γ is the reflection coefficient. This reflection coefficient has components due to discontinuities at both the waveguide-horn junction and the aperture. In a pyramidal horn the E- and H-plane flare discontinuities contribute reactive components to the antenna impedance which are of opposite sign and tend to cancel. A similar situation occurs at the aperture and thus the pyramidal horn is intrinsically well matched. For unmatched standard pyramidal horns, $|\Gamma| < 0.09$ and the effect on the gain is less than 0.04 dB. In sectoral horns, this E- and H-plane cancellation of reflection does not occur. Mismatch losses are larger (see Section 9.5.5) and should be taken into account. They are usually easily obtained from standing wave measurements on the horn.

6.5 Paraboloidal reflectors

6.5.1 Geometrical considerations

A paraboloidal reflector is usually the most convenient way of obtaining high directivity at microwave and millimeter-wave frequencies. To be effective the aperture width of the reflector must be more than about two wavelengths; otherwise a plane or corner reflector is adequate. In Fig. 6.10a, all radiation from f in the direction of the reflecting surface arrives in phase at a plane parallel to the aperture at $z = 0$. For equality of the two paths shown, $2f = \rho(1 + \cos \psi)$ or

$$\rho = \frac{2f}{1 + \cos \psi} \tag{6.36}$$

which is the equation of a parabola. Since $\rho = x/(\sin \psi) = f/\cos^2 (\psi/2)$

$$\frac{x}{2f} = \tan \frac{\psi}{2} \tag{6.37}$$

and for an aperture of diameter d, $x = d/2$ at $\psi = \psi_0$ so that

$$f/d = \frac{1}{4} \cot \frac{\psi_0}{2} \tag{6.38}$$

which relates focal length f, aperture width d and angle subtended by the aperture at the focus $2\psi_0$.

Fields radiated from the focal point f decay as ρ^{-1}; i.e. they are spherical waves, and become plane on reflection. Consequently, fields reflected from the center of the aperture undergo less spatial attenuation than those reflected from extremities. As a result, for an isotropic source at f the field distribution in the aperture ($z = 0$) plane is a maximum at the center and decreases toward the edges by a factor

$$f/\rho = \cos^2 \frac{\psi}{2} = \frac{1}{1 + (x/2f)^2} \tag{6.39}$$

and illumination of the reflector edges $x = \pm d/2$ is reduced by a factor $[1 + (d/4f)^2]^{-1}$. As f/d ranges from 0.25, ($\psi_0 = 90°$) corresponding to a low aperture efficiency, low noise antenna system to $f/d = 0.75$ ($\psi_0 = 37°$) for an antenna with higher aperture efficiency but more 'spillover' and consequently higher antenna noise temperature, this edge illumination factor varies from 0.5 to 0.9.

If the reflector is a parabolic cylinder with a line source feed radiating cylindrical waves from the focus, spatial attenuation affects the aperture field by a factor $(f/\rho)^{1/2}$.

6.5.2 Radiation field and axial gain

Depolarisation of the field radiated from the feed occurs on reflection from the curved paraboloidal surface. Thus a linearly polarised short dipole $Idl\hat{x}$ A.m at the focus $z = z_0$ in Fig. 6.10*b* produces the aperture field (Love, 1978, p. 28).

$$\bar{E} = -\frac{jZ_0 Idl}{4\lambda r} \exp[-jk(2f - z_0)]\{\hat{x}[(1 + \cos\psi) - (1 - \cos\psi)\cos 2\phi']$$

$$-\hat{y}(1 - \cos\psi)\sin 2\phi'\} \tag{6.40}$$

illustrated in Fig. 6.11. Consequently the radiated field has a cross-polarised component. In the two principal planes however, contributions from cross-polarised aperture fields, being equal and opposite from opposite halves of the aperture, cancel. There the radiation pattern can be obtained solely from the x-component of the aperture field. This aperture field is related to the directive gain of a feed antenna $G_f(\psi, \phi')$ by

$$E_x(\rho', \phi') = \left[\frac{2Z_0 P_r}{4\pi} G_f(\psi, \phi')\right]^{1/2} \frac{\exp[-jk(zf - z_0)]}{\rho} \tag{6.41}$$

which in eqn. 3.27 with the substitution $\rho' = 2f\tan(\psi/2)$ gives the radiation field

$$\bar{E}(r, \theta, \phi) = 2f\bar{A}\exp[-jk(zf - z_0)]\left(\frac{2Z_0 P_r}{4\pi}\right)^{1/2}\int_0^{2\pi}\int_0^{\psi_0}[G_f(\psi, \phi')]^{1/2}$$

$$\times \exp[j2kf\tan(\psi/2)\sin\theta\cos(\phi - \phi')]\tan\frac{\psi}{2}d\psi d\phi' \tag{6.42}$$

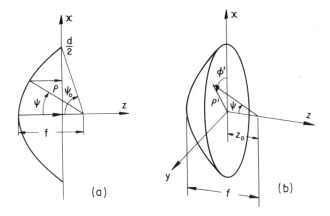

Fig. 6.10 *Coordinates of:*
a Parabolic reflector
b Paraboloidal reflector

AADT – F

The axial ($\theta = 0$) field, for a feed pattern with circular symmetry (independent of ϕ') is

$$\overline{E}(r,0) = j\frac{\hat{\theta}\,4\pi f}{\lambda r}\exp[-jk(r+zf-z_0)]\left(\frac{2Z_0 P_r}{4\pi}\right)^{1/2}\int_0^{\psi_0}$$

$$\times\ [G_f(\psi)]^{1/2}\tan\frac{\psi}{2}d\psi \tag{6.43}$$

The axial gain of the reflector system is

$$G = \frac{\frac{1}{2}Y_0|E_\theta(r,0)|^2}{P_r/(4\pi r^2)} = \frac{4\pi A}{\lambda^2}\cot^2\frac{\psi_0}{2}\left|\int_0^{\psi_0}[G_f(\psi)]^{1/2}\tan\frac{\psi}{2}d\psi\right|^2 \tag{6.44}$$

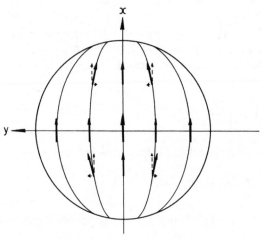

Fig. 6.11   *Aperture field polarisation with an x-polarised feed at the focus of a paraboloidal reflector*

where eqn. 6.38 has been used and $A = \pi d^2/4$ is the aperture area. In eqn. 6.44, $4\pi A/\lambda^2$ is the aperture gain for uniform illumination. The remainder is a gain factor which can be optimized in terms of ψ_0 and $G_f(\psi)$ (Silver, 1949, pp. 425–426), smaller ψ_0 requiring more directive feeds. For typical feeds the maximum value of this gain factor is about 0·83. With the reflector edge illumination 42% say, that of the centre, spillover loss reduces the gain further by a factor of about 0·71 (Thourel, 1960, p. 266), leaving a typical aperture efficiency of 0·59.

Further analysis of paraboloidal reflectors is given elsewhere (e.g. Love, 1978, Wood, 1980).

6.6 Horn reflectors

An offset paraboloidal reflector avoids aperture blockage and most impedance mismatch at the feed. If this feed is a pyramidal horn with its apex at the paraboloid focus and its sides extended to the reflector surface, as in Fig. 6.12, the antenna has little side or back radiation. Thus horn-reflector antennas, with their low side-by-side and back-to-back coupling have been much used in microwave communications. Their low noise temperature was also useful for early satellite ground stations (Crawford *et al.*, 1961) and for the measurement of cosmic background radiation (Penzias and Wilson, 1965).

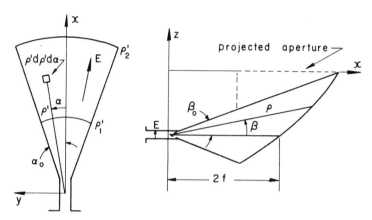

Fig. 6.12   *Coordinates of a rectangular horn-reflector antenna, showing the electric vector E for longitudinal polarisation of the aperture field*

6.6.1 Radiation patterns

With longitudinal polarisation as shown in Fig. 6.12, the aperture electric field is predominantly x-polarised. When fed by a rectangular waveguide supporting only the dominant mode, the electric field distribution in the horn is

$$E_\beta = \cos\left(\frac{\pi\alpha}{2\alpha_0}\right) \frac{\exp(-jk\zeta)}{\zeta} \tag{6.45}$$

where ζ originates at the focus. Reflection from the paraboloidal surface converts this spherical wave into a plane wave with a normalised spatial attenuation factor of, from eqn. 6.36, $2f/\rho = 1 - \sin\beta$. Then the field distribution in the aperture projected on to the $z = 0$ plane is

$$\bar{E}(\alpha,\beta) = \hat{x} E_x(\alpha,\beta) + \hat{y} E_y(\alpha,\beta)$$

$$= (\hat{x}\cos\alpha + \hat{y}\sin\alpha)\cos(\pi\alpha/2\alpha_0)(1 - \sin\beta) \tag{6.46}$$

The far-field pattern of $E_x(\alpha,\beta)$ is given by eqn. 3.1. Since $x = \rho'\cos\alpha, y = \rho'\sin\alpha$, where, from Fig. 6.12, $\rho' = \rho\cos\beta$, this becomes

$$\bar{E}(r,\theta,\phi) = \bar{A} \int_{\rho_1'}^{\rho_2'} \int_{-\alpha_0}^{\alpha_0} E_x(\alpha,\beta) \exp[jk\rho'\sin\theta\cos(\phi-\alpha)] \rho'd\rho'd\alpha \tag{6.47}$$

or, with $\rho' = 2f\cos\beta/(1 - \sin\beta)$,

$$\bar{E}(r,\theta,\phi) = \bar{A}4f^2 \int_{-\alpha_0}^{\alpha_0} \int_{-\beta_0}^{\beta_0} E_x(\alpha,\beta) \frac{\cos\beta}{(1-\sin\beta)^2}$$

$$\times \exp\left[j2kf \frac{\cos\beta}{1-\sin\beta} \sin\theta \cos(\phi-\alpha)\right] d\alpha d\beta \qquad (6.48)$$

Using eqn. 6.46, eqn. 6.48 gives, in the longitudinal or $x-z$ plane,

$$\bar{E}(r,\theta,0) = \frac{\hat{\theta}j \exp(-jkr)}{\lambda r}(1+\cos\theta)2f^2 \int_{-\beta_0}^{\beta_0} \int_{-\alpha_0}^{\alpha_0} \cos\left(\frac{\pi\alpha}{2\alpha_0}\right)$$

$$\times \cos\alpha \frac{\cos\beta}{1-\sin\beta} \exp\left(j2kf \frac{\cos\beta}{1-\sin\beta} \sin\theta \cos\alpha\right) d\alpha d\beta \qquad (6.49)$$

which can be computed without difficulty for the main beam and first few side-lobes (Crawford *et al.*, 1961). For larger angles, Siller (1975) used a method which avoids double numerical integration. The pattern resembles that of a uniform distri-bution but with a slightly broader beam and lower sidelobe levels due to the vari-ation of $E(\alpha,\beta)$ by a ratio $(1+\sin\beta_0)/(1-\sin\beta_0)$ across the aperture (eqn. 6.46).

$\bar{E}(r,\theta,\pi/2)$, the pattern in the $y-z$ plane, is obtained from eqn. 6.49 by replacing $\hat{\theta}$ by $-\hat{\phi}$ and $\cos\alpha$ by $\sin\alpha$ in the exponential only. The pattern is somewhat broader and with lower first sidelobe levels than that of a cosinusoidal distribution, as anticipated from the factor $\cos\alpha$ in eqn. 6.46.

The pattern of the cross-polarised components $E_y(\alpha,\beta)$ may be obtained through eqn. 2.48. In the $x-z$ plane, contributions from opposite halves of the aperture, being equal and opposite, cancel. In the $y-z$ plane, the cross-polarised pattern $\bar{E}(r,\theta,\pi/2)$ is given by the right-hand side of eqn. 6.49 with $\sin\alpha$ replacing $\cos\alpha$ throughout the integrand. This pattern has an axial null and its main lobes are in the direction of the first nulls of the $\phi=0$ principal plane pattern of $E_x(\alpha,\beta)$, but well over 20 dB below the peak of its main beam.

If the waveguide feed in Fig. 6.12 is rotated $90°$, transversal polarisation of the aperture field results. The aperture field is then

$$\bar{E}(\alpha,\beta) = (\hat{x}\sin\alpha + \hat{y}\cos\alpha)\cos\left(\frac{\pi\beta}{2\beta_0}\right)(1-\sin\beta) \qquad (6.50)$$

and predominantly y-polarised. The radiation pattern of the y-component of eqn. 6.50 in the transverse or $y-z$ plane is

$$\bar{E}(r,\theta,\pi/2) = \frac{\hat{\theta}j \exp(-jkr)}{\lambda r}(1+\cos\theta)2f^2 \int_{-\beta_0}^{\beta_0} \int_{-\alpha_0}^{\alpha_0} \cos\left(\frac{\pi\beta}{2\beta_0}\right)$$

$$\times \cos\alpha \frac{\cos\beta}{1-\sin\beta} \exp\left(j2kf \frac{\cos\beta}{1-\sin\beta} \cos\alpha \sin\theta\right) d\alpha d\beta \qquad (6.51)$$

and the other principal plane patterns can be obtained accordingly. The pattern of eqn. 6.51 is slightly less directive than eqn. 6.49.

With a square waveguide feed exciting equally longitudinally and transversely polarised aperture fields in phase quadrature, circular polarisation results. These circularly polarised radiation patterns can be calculated from linear combinations of the patterns for longitudinal and transverse polarisations and are less directive than them. Reflection depolarisation also produces lobes of undesired circular polarisation off the beam axis (Crawford *et al.*, 1961).

6.6.2 Horn-reflector gain

Integration of eqn. 6.49 with $\theta = 0$ yields the axial electric field for longitudinal polarisation

$$E_\theta(r,0) = \frac{j\exp(-jkr)}{\lambda r}8f^2\ln\left(\frac{1+\sin\beta_0}{1-\sin\beta_0}\right)\frac{\pi/2\alpha_0}{(\pi/2\alpha_0)^2-1}\cos\alpha_0 \tag{6.52}$$

The power radiated through the aperture is

$$P_r = 2Y_0f^2\int_{-\alpha_0}^{\alpha_0}\int_{-\beta_0}^{\beta_0}(|E_x|^2+|E_y|^2)\frac{\cos\beta}{(1-\sin\beta)^2}d\alpha d\beta \tag{6.53}$$

$$= 4Y_0f^2\alpha_0\sin\beta_0 \tag{6.54}$$

when eqn. 6.46 is used. Eqns. 6.52 and 6.54 give for the axial gain with longitudinal polarisation

$$G = \frac{\frac{1}{2}Y_0|E_\theta(r,0)|^2}{P_r/4\pi r^2}$$

$$= \frac{32\pi f^2}{\lambda^2}\left[\ln\left(\frac{1+\sin\beta_0}{1-\sin\beta_0}\right)\frac{\pi/2\alpha_0}{(\pi/2\alpha_0)^2-1}\right]^2\frac{\cos^2\alpha_0}{\alpha_0\sin\beta_0} \tag{6.55}$$

As an example, a horn-reflector antenna for microwave communications has a horn with flare angles $\alpha_0 = \beta_0 = 20°$ and a reflector with a focal length $f = 1.855$ m. Its gain from eqn. 6.55 is $G = 71.03/\lambda^2$ with λ in meters. At 6 GHz, $\lambda = 0.05$ m and the gain is 44.53 dB. Since the aperture area is

$$A = 4f^2\int_{-\beta_0}^{\beta_0}\int_{-\alpha_0}^{\alpha_0}\frac{\cos\beta}{(1-\sin\beta)^2}d\alpha d\beta = 16f^2\alpha_0\tan\beta_0\sec\beta_0$$

the aperture efficiency is

$$\frac{A_e}{A} = \frac{1}{2}\left[\frac{\cos\alpha_0}{\alpha_0\tan\beta_0}\ln\left(\frac{1+\sin\beta_0}{1-\sin\beta_0}\right)\frac{\pi/2\alpha_0}{(\pi/2\alpha_0)^2-1}\right]^2 \tag{6.56}$$

or 0.76 for the above parameters. This efficiency is substantially higher than for most paraboloidal reflectors. It decreases as α_0 and β_0 increase, as shown in Fig. 6.13. For small α_0 and β_0 the aperture efficiency approaches 0.81, the value in Table 5.2 for an in-phase uniform and cosinusoidal distribution.

For transverse polarisation, the axial field is, from eqn. 6.51 with $\theta = 0$,

$$E_\theta(r,0) = j\frac{\exp(-jkr)}{\lambda r}4f^2\sin\alpha_0 I \tag{6.57}$$

where

$$I = \int_{-\beta_0}^{\beta_0} \frac{\cos\beta}{1-\sin\beta}\cos\left(\frac{\pi\beta}{2\beta_0}\right)d\beta \tag{6.58}$$

The power radiated is, from eqns. 6.50 and 6.53,

$$P_r = 4Y_0 f^2 \alpha_0 \sin\beta_0 \frac{(\pi/\beta_0)^2}{(\pi/\beta_0)^2 - 1} \tag{6.59}$$

on the on-axis gain

$$G = \frac{32\pi f^2}{\lambda^2} \frac{\sin^2\alpha_0}{\alpha_0 \sin\beta_0}\left[1 - \left(\frac{\beta_0}{\pi}\right)^2\right]I^2 \tag{6.60}$$

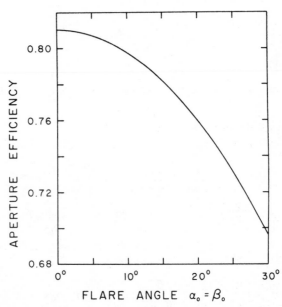

Fig. 6.13 *Aperture efficiency of a square horn-reflector antenna against flare angle*

With $\alpha_0 = \beta_0 = 20°$ and $f = 1\cdot855\,\text{m}$, $I = 0\cdot4497$ and $G = 67\cdot69/\lambda^2$ with λ in meters. This yields an effective area of $5\cdot33\,\text{m}^2$ or an aperture efficiency of $0\cdot72$ and a gain at 6 GHz of $44\cdot33$ dB.

The measured gain may be expected to be lower than these predicted values by possibly $0\cdot1$ dB or less, for an accurately made antenna. Spillover over the outer edge of the reflector would be responsible for much of this loss (Crawford *et al.*, 1961). Measured horn-reflector gain increases with frequency monotonically, as predicted, deviating only as the surface or structural imperfections become substantial in terms of the wavelength (Jull and Deloli, 1964). There is no evidence of an oscillation in the gain against wavelength such as is evident for smaller horns in Figs. 6.7 and 6.9. Mismatch losses measured in a horn reflector of the above dimensions are about 0·02 dB, mainly due to the horn-waveguide junction.

The axial near-field gain reduction of a horn-reflector antenna is very similar to that of a uniform and cosinusoidal distribution in a square aperture of the same area (Jull and Deloli, 1964). Its value at $r = a^2/\lambda$, for example, is 0·923 (cf. Table 5.2).

Chapter 7

Diffraction by conductors with sharp edges

7.1 Boundary conditions on conductors and at edges

The preceding chapters use the approximate Kirchhoff theory of diffraction. As indicated in Sections 2.6 and 6.1 it does not, in general, satisfy the boundary conditions. This chapter gives a few elementary diffraction solutions for conductors with edges in which boundary conditions are rigorously satisfied. These exact solutions are the basis of the newer approximate method of aperture antenna analysis used in the remaining chapters.

When an electromagnetic wave with fields $\bar{E}^i$, $\bar{H}^i$ is incident on a perfectly conducting surface, a current is excited and the fields $\bar{E}^s$, $\bar{H}^s$ are scattered. The total electric and magnetic fields are

$$\bar{E} = \bar{E}^i + \bar{E}^s$$
$$\bar{H} = \bar{H}^i + \bar{H}^s \tag{7.1}$$

and the boundary conditions on the surface are $\hat{n} \cdot \bar{H} = 0$ and

$$\hat{n} \times \bar{E} = 0$$
$$\hat{n} \times \bar{H} = \bar{J} \tag{7.2}$$

where $\hat{n}$ is a unit vector outwardly normal to the surface. $\bar{J}$ is the surface current density, or, from eqn. 7.2, the discontinuity of the tangential magnetic field across the surface.

The mathematically convenient concept of an infinitely thin, perfectly conducting screen, practical at radio frequencies, will be used. In the aperture of such a screen $\bar{J} = 0$ and if the conductor is planar the scattered fields $\bar{E}^s$ are excited by currents wholly tangential to the plane. Then the boundary conditions in the aperture, which follow from eqn. 7.2, are

$$\hat{n} \times \bar{H}^s = 0$$
$$\hat{n} \cdot \bar{E}^s = 0 \tag{7.3}$$

Solutions of electromagnetic scattering problems for conductors with edges may satisfy Maxwell's equations through eqn. 2.4, the boundary conditions of eqn. 7.2 and behave in the proper way at large distances from the scattering object, yet not be unique (Bouwkamp, 1954). For uniqueness, the field behaviour near sharp edges must be specified. Although some field components become infinite at an edge, the order of this singularity is restricted by the requirement that the energy stored in the vicinity of the edge must be finite, i.e. the edge must not behave as a

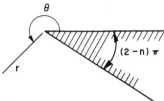

Fig. 7.1 *Coordinates of a wedge*

source. For the two-dimensional conducting wedge of Fig. 7.1, it has been shown that this leads to the following behaviour of the field components near the edge (Meixner, 1954)

$$\left. \begin{array}{l} E_z, H_z \ = \ 0(r^{\alpha+1}) \\ H_r, H_\theta, E_r, E_\theta \ = \ 0(r^\alpha) \end{array} \right\} \ r \to 0 \tag{7.4}$$

with
$$\alpha = \frac{\pi}{2\pi - \phi} - 1 \tag{7.5}$$

When $\phi = 0$, corresponding to a conducting half-plane in $y = 0, x > 0, \alpha = -1/2$ and the fields parallel to the edge are finite whereas those normal to it behave as $r^{-1/2}$ as $r \to 0$ (Bouwkamp, 1946). For a right-angled conducting wedge ($\phi = \pi/2$), the order of this singularity is $\alpha = -1/3$ and for a plane ($\phi = \pi$) it is of course zero.

7.2 Babinet's principle

Babinet's principle relates the diffracted field produced by a conducting disc of arbitrary shape to the diffracted field of an aperture of the same shape in a conducting screen. The diffracting obstacles are complementary in the sense that, fitted together, they form a plane conducting surface of infinite extent. If the disc is in the $y = 0$ plane and excited by a disturbance in $y > 0$, the light intensity at a point in $y < 0$ plus the intensity at that point when the aperture replaces the disc in $y = 0$ gives, according to the optical form of Babinet's principle, the intensity which would exist there if neither diffracting obstacle were present.

For an electromagnetic Babinet's principle let $\bar{E}^i$, $\bar{H}^i$ be the fields produced by the source with no diffraction and $\bar{E}^s$, $\bar{H}^s$ be the scattered fields as in eqn. 7.1. Let the aperture in a plane screen be designated A and the remainder of the conducting screen S. With subscripts t and n denoting tangential and normal field components, the first boundary condition of eqn. 7.2 and the second of eqn. 7.3 give, respectively,

$$\bar{E}^i_t + \bar{E}^s_t = 0 \qquad \text{on S} \tag{7.6}$$

$$E^s_n = 0 \qquad \text{on A} \tag{7.7}$$

The screen is now replaced by its complement in which A is the conducting region and S the aperture region. The total fields are now $\bar{E}'$, $\bar{H}'$ and the boundary conditions yield

$$H'_n = 0 \qquad \text{on A} \tag{7.8}$$

$$\bar{H}'_t - \bar{H}^{i'}_t = 0 \qquad \text{on S} \tag{7.9}$$

If the two fields are related by $\bar{H}^{i'} = Y_0 \bar{E}^i$ and this is inserted in eqn. 7.9, then in $y < 0$ $\bar{H}'_t = Y_0 \bar{E}_t = -Y_0 \bar{E}^s_t$ from eqn. 7.6. Also, comparison of eqns. 7.7 and 7.6 leads to $\bar{H}' = -Y_0 \bar{E}^s$ in $y < 0$, or with eqn. 7.1,

$$\bar{E} + Z_0 \bar{H}' = \bar{E}^i \tag{7.10}$$

Unlike its optical form, this electromagnetic form of Babinet's principle, by Booker (1946), specifies field polarisation. It may be used to obtain the fields $\bar{E}'$, $\bar{H}'$ in the presence of a plane conducting obstacle from known diffraction solutions for the fields $\bar{H}$, $\bar{E}$, respectively, of its Babinet complement. An example is given in Section 7.4.2.

7.3 Two-dimensional electromagnetic fields

Electromagnetic problems in which the fields are independent of one Cartesian coordinate are essentially scalar. If all fields are uniform in the z-direction, say, spatial derivatives in z vanish and Maxwell's equations divide into two independent sets. In free space and with a time dependence $\exp(j\omega t)$, these are

$$\frac{\partial E_z}{\partial y} = -j\omega\mu_0 H_x, \quad \frac{\partial E_z}{\partial x} = j\omega\mu_0 H_y, \quad \frac{\partial H_y}{\partial x} - \frac{\partial H_x}{\partial y} = j\omega\epsilon_0 H_z \tag{7.11}$$

$$\frac{\partial H_z}{\partial y} = j\omega\epsilon_0 E_x, \quad \frac{\partial H_z}{\partial x} = -j\omega\epsilon_0 E_y, \quad \frac{\partial E_y}{\partial x} - \frac{\partial E_x}{\partial y} = -j\omega\mu_0 H_z \tag{7.12}$$

The first set, consisting of E_z, H_x, H_y, is called a transverse electric (TE) polarised field. The other set is called transverse magnetic (TM) polarised. Any two-dimensional field may be so resolved. The solution in a diffraction problem for TE polarisation must satisfy the scalar wave equation for E_z obtained from eqn. 7.11:

$$\left(\frac{\partial^2}{\partial x^2} + \frac{\partial^2}{\partial y^2} + k^2 \right) E_z = 0 \tag{7.13}$$

and the boundary conditions. For TM polarised fields, which scatter independently of TE polarised fields from conducting obstacles uniform in the z-direction, H_z replaces E_z in eqn. 7.13.

7.4 Diffraction by a conducting half-plane

7.4.1 TE polarisation

The exact solution for plane wave diffraction by a conducting half-plane was first obtained by Sommerfeld (1896) by his method of many-valued wave functions. Integral equation solutions now provide a more direct route to the exact solution. The procedure used here is that given by Clemmow (1951) and others.

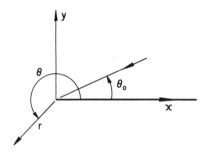

Fig. 7.2   *Diffraction by a conducting half-plane*

A thin, perfectly conducting screen lies in $y = 0$, $x > 0$ of Fig. 7.2 and an arbitrarily polarised electromagnetic plane wave is normally incident upon the edge, but at an angle of incidence $0 < \theta_0 < \pi$ to the conducting plane. Oblique incidence is considered later. The TE polarised incident field is

$$E_z^i = \exp[jk(x \cos \theta_0 + y \sin \theta_0)] = \exp[jkr \cos(\theta - \theta_0)] \qquad (7.14)$$

The scattered fields may be written as a spectrum of outgoing plane waves, as in Section 2.4,

$$E_z^s = \int_{-\infty}^{\infty} F(k_x) \exp[-j(k_x x \pm k_y y)] \, dk_x, \qquad (7.15)$$

$$H_{x,\pm}^s = \pm \frac{1}{\omega \mu_0} \int_{-\infty}^{\infty} k_y F(k_x) \exp[-j(k_x x \pm k_y y)] \, dk_x \qquad (7.16)$$

$$H_{y,\pm}^s = -\frac{1}{\omega \mu_0} \int_{-\infty}^{\infty} k_x F(k_x) \exp[-j(k_x x + k_y y)] \, dk_x \qquad (7.17)$$

where the upper signs apply in $y \geqslant 0$ and the lower signs in $y \leqslant 0$. Here $k_x = k \cos \alpha$, where α is the angle of the scattered wave from the screen. α may be real ($k_x^2 < k^2$) or complex ($k_x^2 \geqslant k^2$) corresponding to propagating or evanescent scattered fields. For these cases, respectively,

$$k_y = \sqrt{k^2 - k_x^2}$$
$$= -j \sqrt{k_x^2 - k^2} \tag{7.18}$$

and the integration contour, shown in Fig. 7.3, avoids the branch points at $k_x = \pm k$. It is indented above a pole at $k_x = -k_0 = -k \cos \theta_0$ for reasons apparent in the integral equation solution.

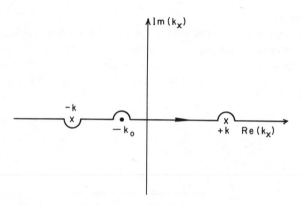

Fig. 7.3 *Integration contour in the complex k_x-plane*

The boundary conditions in $y = 0$ corresponding to eqn. 7.2 are

$$E_z^i + E_z^s = 0, \qquad x > 0 \tag{7.19}$$

$$H_{x,+}^s = H_{x,-}^s, \qquad x < 0 \tag{7.20}$$

which, with eqns. 7.14–7.16 give the integral equations

$$\int_{-\infty}^{\infty} F(k_x) \exp(-jk_x x) dk_x = -\exp(jk_0 x), \qquad x > 0 \tag{7.21}$$

$$\int_{-\infty}^{\infty} k_y F(k_x) \exp(-jk_x x) dk_x = 0, \qquad x < 0 \tag{7.22}$$

These dual integral equations are solved by the techniques of contour integration in Appendix A.4, giving for the spectrum of the scattered field:

$$F(k_x) = \frac{1}{2\pi j} \sqrt{\frac{k - k_0}{k + k_x}} \frac{1}{(k_x + k_0)} \tag{7.23}$$

and consequently the scattered field is, from eqn. 7.15,

$$E_z^s = \frac{\sqrt{k - k_0}}{2\pi j} \int_{-\infty}^{\infty} \frac{1}{(k_x + k_0)\sqrt{k_x + k}} \exp[-j(k_x x \pm k_y y)] dk_x \tag{7.24}$$

A procedure for reducing this integral to complex Fresnel integrals has been given by Copson (1946), by Noble (1958, p. 33), by Clemmow (1966) and in Born and Wolf (1964, p. 368). Clemmow's procedure is outlined in Appendix A.5. The total field in cylindrical coordinates r, θ, valid for $0 < \theta < 2\pi$, is

$$E_z = \frac{\exp(j\pi/4)}{\sqrt{\pi}} \left\{ \exp[jkr\cos(\theta - \theta_0)] F\left[-\sqrt{2kr}\cos\left(\frac{\theta - \theta_0}{2}\right)\right] \right.$$

$$\left. - \exp[jkr\cos(\theta + \theta_0)] F\left[-\sqrt{2kr}\cos\left(\frac{\theta + \theta_0}{2}\right)\right] \right\} \qquad (7.25)$$

where

$$F(a) = \int_a^\infty \exp(-j\tau^2) d\tau \qquad (7.26)$$

is the complex Fresnel integral. In eqn. 7.25,

$$F(a) + F(-a) = \int_{-\infty}^\infty \exp(-j\tau^2) d\tau = \sqrt{\pi} \exp(-j\pi/4) \qquad (7.27)$$

has been used to include the incident field of eqn. 7.14. The magnetic field components may be obtained from eqn. 7.25 by differentiation

$$H_\theta = \frac{1}{j\omega\mu_0} \frac{\partial E_z}{\partial r}, \quad H_r = -\frac{1}{j\omega\mu_0 r} \frac{\partial E_z}{\partial \theta} \qquad (7.28)$$

7.4.2 TM polarisation

The solution for TM-polarised plane wave diffraction by a conducting half-plane may be obtained in a manner similar to the above, but it is simpler to use Babinet's principle. A TM-polarised wave

$$H_z^i = \exp[jkr\cos(\theta - \theta_0)] \qquad (7.29)$$

is incident at an angle $0 < \theta_0 < \pi$ on a conducting half-plane in $y = 0$, $-\infty < x < 0$. The total magnetic field behind the screen ($\pi < \theta < 2\pi$) is, from eqn. 7.10, given by $E_z^i - E_z$, where E_z^i is given by eqn. 7.14 and E_z by eqn. 7.25. Hence, using eqn. 7.25, the total magnetic field is

$$H_z = \frac{\exp(+j\pi/4)}{\sqrt{\pi}} \left\{ \exp[jkr\cos(\theta - \theta_0)] F\left[\sqrt{2kr}\cos\left(\frac{\theta - \theta_0}{2}\right)\right] \right.$$

$$\left. + \exp[jkr\cos(\theta + \theta_0)] F\left[-\sqrt{2kr}\cos\left(\frac{\theta + \theta_0}{2}\right)\right] \right\} \qquad (7.30)$$

in the coordinates of Fig. 7.4. For the angular coordinates to conform with those of Fig. 7.2, θ_0 is replaced by $\theta_0 + \pi$ and θ by $\theta - \pi$. Then the sign of the argument of the first Fresnel integral changes sign, i.e.

$$H_z = \frac{\exp(+j\pi/4)}{\sqrt{\pi}} \left\{ \exp[jkr\cos(\theta - \theta_0)] F\left[-\sqrt{2kr}\cos\left(\frac{\theta - \theta_0}{2}\right)\right] \right.$$

$$\left. + \exp[jkr\cos(\theta + \theta_0)] F\left[-\sqrt{2kr}\cos\left(\frac{\theta + \theta_0}{2}\right)\right] \right\} \qquad (7.31)$$

in the coordinates of Fig. 7.2. This result is valid for all θ, θ_0 in the range $(0, 2\pi)$.

Notice that in the aperture ($\theta = \pi$) the total tangential magnetic field reduces to the incident field of eqn. 7.29 in accordance with eqn. 7.3. This also occurs for TE-polarisation; there is no scattered tangential magnetic field in the aperture. The scattered tangential electric fields are non-zero in the aperture, however.

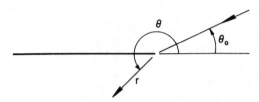

Fig. 7.4   *Coordinates for diffraction by a half-plane complementary to that of Fig. 7.2*

The TM-polarised electric field components are obtained from eqn. 7.31 by

$$E_\theta = -\frac{1}{j\omega\epsilon_0}\frac{\partial H_z}{\partial r}, \quad E_r = \frac{1}{j\omega\epsilon_0 r}\frac{\partial H_z}{\partial\theta} \tag{7.32}$$

7.5 Oblique incidence on a half-plane

An arbitrarily polarised electromagnetic wave is obliquely incident on a conducting half-plane in $y = 0, x > 0$ in Fig. 7.5. The phase of the incident wave is

$$\exp(jk\zeta_0) = \exp[jk(x\cos\theta_0\sin\beta_0 + y\sin\theta_0\sin\beta_0 + z\cos\beta_0)] \tag{7.33}$$

which is got from the two-dimensional form ($\beta_0 = \pi/2$) by replacing k by $k\sin\beta_0$ and multiplying by $\exp(jkz\cos\beta_0)$. The scattered fields will also vary with z as

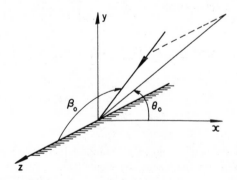

Fig. 7.5   *Oblique incidence of a plane wave on a half-plane*

$\exp(jkz\cos\beta_0)$ so on expanding Maxwell's equations $\partial/\partial z$ may be replaced with $jk\cos\beta_0$. As in the two-dimensional case, the equations may be split into two sets,

TE-polarised fields ($H_z = 0$) and TM-polarised fields ($E_z = 0$). On setting $H_z = 0$ and eliminating H_x, H_y, E_x and E_y one obtains

$$\frac{\partial^2 E_z}{\partial x^2} + \frac{\partial^2 E_z}{\partial y^2} + k^2 \sin^2\beta_0 E_z = 0 \tag{7.34}$$

and the field components are related by

$$\hat{x}E_x + \hat{y}E_y = j\frac{\cos\beta_0}{k\sin^2\beta_0}\left(\hat{x}\frac{\partial E_z}{\partial x} + \hat{y}\frac{\partial E_z}{\partial y}\right) \tag{7.35}$$

$$\hat{x}H_x + \hat{y}H_y = \frac{1}{-j\omega\mu_0 \sin^2\beta_0}\left(\hat{x}\frac{\partial E_z}{\partial y} - \hat{y}\frac{\partial E_z}{\partial x}\right) \tag{7.36}$$

The boundary conditions are those of eqn. 7.2. The problem therefore is identical to that of Section 7.4.1 if $k \sin\beta_0$ replaces k and all fields are multiplied by $\exp(jkz \cos\beta_0)$. It is evident that this procedure applied to any two-dimensional solution of the wave equation yields a solution of the three-dimensional wave equation

$$\frac{\partial^2 E_z}{\partial x^2} + \frac{\partial^2 E_z}{\partial y^2} + \frac{\partial^2 E_z}{\partial z^2} + k^2 E_z = 0$$

when the diffracting obstacle is two-dimensional and perfectly conducting so no coupling between TE- and TM- polarised fields occurs on diffraction.

If $\sin\beta_0$ times eqn. 7.33 is the incident field E_z^i, from eqns. 7.35 and 7.36 the incident wave is

$$\bar{E}^i = (-\hat{x}\cos\theta_0 \cos\beta_0 - \hat{y}\sin\theta_0 \cos\beta_0 + \hat{z}\sin\beta_0)\exp(jk\zeta_0) \tag{7.37}$$

$$\bar{H}^i = Y_0(-\hat{x}\sin\theta_0 + \hat{y}\cos\theta_0)\exp(jk\zeta_0) \tag{7.38}$$

The total E_z field is, from eqn 7.25,

$$E_z = \sin\beta_0 \frac{\exp(jkz\cos\beta_0 + j\pi/4)}{\sqrt{\pi}}\left\{\exp[jkr\sin\beta_0 \cos(\theta - \theta_0)]\right.$$

$$\times F\left[-\sqrt{2kr\sin\beta_0}\cos\left(\frac{\theta - \theta_0}{2}\right)\right] - \exp[jkr\sin\beta_0 \cos(\theta + \theta_0)]$$

$$\left.\times F\left[-\sqrt{2kr\sin\beta_0}\cos\left(\frac{\theta + \theta_0}{2}\right)\right]\right\} \tag{7.39}$$

and the remaining field components follow from eqns. 7.35 and 7.36.

The TM-polarised fields ($E_z = 0$) may be handled similarly. For these the magnetic field components are related as the electric fields are in eqn. 7.35 and the electric fields are

$$\hat{x}E_x + \hat{y}E_y = \frac{1}{j\omega\epsilon_0 \sin^2\beta_0}\left(\hat{x}\frac{\partial H_z}{\partial y} - \hat{y}\frac{\partial H_z}{\partial x}\right) \tag{7.40}$$

7.6 Line source diffraction by a half-plane

Suppose the field incident on a conducting half-plane, instead of being a plane wave from a distant source, is from a local source. The exact solution for TE- and TM-polarised line sources, sources uniform in the direction parallel to the edge, can be obtained by an integration over all angles θ_0 of the corresponding plane wave solution. Clemmow (1950, 1966, pp. 92–97) used this procedure (see also Born and Wolf, 1964, pp. 580–584). For a TE-polarised or electric line source,

$$E_z^i = \sqrt{\frac{\pi}{2}} \exp(-j\pi/4) H_0^{(2)}(kR) \simeq \frac{\exp(-jkR)}{\sqrt{kR}} \tag{7.41}$$

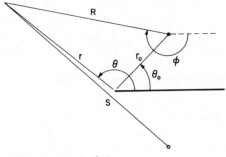

Fig. 7.6   *Line source diffraction by a half-plane*

located at r_0, θ_0 from the edge of the half-plane in Fig. 7.6, the total field at r, θ is

$$E_z = \sqrt{\frac{2}{\pi}} \exp(j\pi/4) \left\{ \exp(-jkR) \int_{-p}^{\infty} \frac{\exp(-j\tau^2)d\tau}{\sqrt{\tau^2 + 2kR}} \right.$$
$$\left. - \exp(-jkS) \int_{-q}^{\infty} \frac{\exp(-j\tau^2)d\tau}{\sqrt{\tau^2 + 2kS}} \right\} \tag{7.42}$$

where

$$p = 2\sqrt{\frac{krr_0}{r_0 + r + R}} \cos\left(\frac{\theta - \theta_0}{2}\right)$$
$$q = 2\sqrt{\frac{krr_0}{r_0 + r + S}} \cos\left(\frac{\theta + \theta_0}{2}\right) \tag{7.43}$$

R and S are the distances from the source and the image, respectively, to the field point, r, θ, i.e.

$$R^2 = r^2 + r_0^2 - 2rr_0 \cos(\theta - \theta_0)$$
$$S^2 = r^2 + r_0^2 - 2rr_0 \cos(\theta + \theta_0) \tag{7.44}$$

If $k(r + r_0) \gg 1$, eqn. 7.42 becomes

$$E_z = \sqrt{\frac{2}{\pi}} \exp(j\pi/4) \left\{ \frac{\exp(-jkR)}{\sqrt{k(r+r_0+R)}} F[-p] - \frac{\exp(-jkS)}{\sqrt{k(r+r_0+S)}} F[-q] \right\}$$

(7.45)

where $F(a)$ is the complex Fresnel integral defined in eqn. 7.26. Further simplification results when $r_0 \gg r$. Then eqn. 7.44 becomes

$$R \approx r_0 - r \cos(\theta - \theta_0)$$ (7.46)

$$S \approx r_0 - r \cos(\theta + \theta_0)$$

and

$$E_z = \frac{\exp(-jkr_0 + j\pi/4)}{\sqrt{\pi k r_0}} \left\{ \exp[jkr \cos(\theta - \theta_0)] F\left[-\sqrt{2kr} \cos\left(\frac{\theta - \theta_0}{2} \right) \right] \right.$$

$$\left. - \exp[jkr \cos(\theta + \theta_0)] F\left[-\sqrt{2kr} \cos\left(\frac{\theta + \theta_0}{2} \right) \right] \right\}$$

(7.47)

which is the solution for plane wave diffraction by a half-plane of eqn. 7.25 but with a distant cylindrical wave.

Similarly, the distant field of a line source near the edge can be obtained from eqn. 7.45 by letting $r \gg r_0$ in eqn. 7.45, which yields eqn. 7.47 with r and r_0 interchanged. This result, which is exact, can also be obtained directly from eqn. 7.25 by reciprocity.

In Fig. 7.6, the far field at r, θ due to an electric line source at r_0, θ_0 parallel to the edge of a conducting half-plane in $y = 0, x > 0$, is a cylindrical wave of the form

$$E_z(r, \theta, r_0, \theta_0) = \frac{\exp(-jkr)}{\sqrt{kr}} f(\theta, r_0, \theta_0)$$

(7.48)

The reciprocity theorem states that if source and observer are interchanged the same field is observed. Hence the field at r_0, θ_0 due to a line source at r, θ far from the edge of the half-plane is

$$E_z(r_0, \theta_0, r, \theta) = \frac{\exp(-jkr_0)}{\sqrt{kr_0}} f(\theta_0, r, \theta)$$

(7.49)

Since the field striking the edge is essentially plane, we may use eqn. 7.25 and

$$f(\theta_0, r, \theta) = \frac{\exp(j\pi/4)}{\sqrt{\pi}} \left\{ \exp[jkr \cos(\theta - \theta_0) F\left[-\sqrt{2kr} \cos\left(\frac{\theta - \theta_0}{2} \right) \right] \right.$$

$$\left. - \exp[jkr \cos(\theta + \theta_0)] F\left[-\sqrt{2kr} \cos\left(\frac{\theta + \theta_0}{2} \right) \right] \right\}$$

(7.50)

Hence the far field of the line source is

$$E_z(r, \theta, r_0, \theta_0) = \frac{\exp[-j(kr - \pi/4)]}{\sqrt{\pi kr}} \left\{ \exp[jkr_0 \cos(\theta - \theta_0)] \right.$$

$$\times F\left[-\sqrt{2kr_0} \cos\left(\frac{\theta - \theta_0}{2} \right) \right] - \exp[jkr_0 \cos(\theta + \theta_0)]$$

$$\times F\left[-\sqrt{2kr_0}\,\cos\left(\frac{\theta + \theta_0}{2}\right)\right]\Bigg\}\tag{7.51}$$

Solutions for TM-polarised or magnetic line sources are given by the above with E_z replaced by H_z and the $(-)$ sign between the integrals of eqns. 7.42, 7.45, 7.47, 7.50 and 7.51 replaced by $(+)$.

7.7 Diffraction by a wedge

7.7.1 Series solution

A solution for plane wave diffraction by a two-dimensional wedge cannot be obtained in the manner of Section 7.4.1. Instead, a series solution is given. For a line source at r_0, θ_0 from the conducting wedge of Fig. 7.7, the total TE-polarised electric field at r, θ is

$$E_z = \frac{2}{n}\sqrt{\frac{\pi}{2j}}\sum_{m=0}^{\infty}\epsilon_m J_{m/n}(kr)H^{(2)}_{m/n}(kr_0)\sin\left(\frac{m\theta_0}{n}\right)\sin\left(\frac{m\theta}{n}\right),\quad r<r_0 \tag{7.52}$$

$$= \frac{2}{n}\sqrt{\frac{\pi}{2j}}\sum_{m=0}^{\infty}\epsilon_m J_{m/n}(kr_0)H^{(2)}_{m/n}(kr)\sin\left(\frac{m\theta_0}{n}\right)\sin\left(\frac{m\theta}{n}\right),\quad r>r_0 \tag{7.53}$$

where $\epsilon_m = 1$ for $m = 0$ and $\epsilon_m = 2$ for $m > 0$. This result satisfies the boundary conditions $E_z = 0$ on the wedge surfaces $\theta = 0$ and $\theta_0 = 2\pi - \phi = n\pi$. For $r \to 0$, the small argument form of $J_{m/n}(kr)$ gives the correct form of E_z in accordance with eqn. 7.4. The above expressions also satisfy reciprocity and include the incident field of eqn. 7.41.

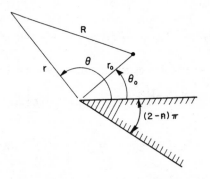

Fig. 7.7   *Line source diffraction by a wedge*

The far-field pattern is obtained by using the asymptotic form of the Hankel function in eqn. 7.53 (see Appendix A.3). Then

$$E_z \simeq \frac{\exp(-jkr)}{\sqrt{kr}} \frac{2}{n} \sum_{m=0}^{\infty} \epsilon_m \exp[j(m\pi/2n)] \, J_{m/n}(kr_0) \, \sin\left(\frac{m\theta_0}{n}\right)$$
$$\times \sin\left(\frac{m\theta}{n}\right) \qquad (7.54)$$

which converges rapidly for $kr_0 \ll 1$. Some numerical values have been given by Wait (1953).

For the plane wave of eqn. 7.14 incident on the conducting wedge of Fig. 7.1, the total TE-polarised electric field is

$$E_z = \frac{2}{n} \sum_{m=0}^{\infty} \epsilon_m \exp[j(m\pi/2n)] J_{m/n}(kr) \sin\left(\frac{m\theta_0}{n}\right) \sin\left(\frac{m\theta}{n}\right) \qquad (7.55)$$

For $n = 2$, the wedge becomes a half-plane and the result is a series form of half-plane solution equivalent to eqn. 7.25.

Corresponding results for TM-polarisation are obtained by replacing E_z by H_z and $\sin(m\theta_0/n) \sin(m\theta/n)$ by $\cos(m\theta_0/n) \cos(m\theta/n)$ in the above expressions. Integral forms of these solutions with a few numerical examples are given by Bowman *et al.* (1969). For derivations of the above expressions see for example, Harrington (1961, pp. 238–242) or James (1974, pp. 63–65).

7.7.2 An asymptotic solution

In the following chapters, a solution for diffraction by a wedge is required in which incident, reflected and diffracted fields are isolated and calculated conveniently. With a half-plane, closed form solutions in terms of Fresnel integrals as in Section 7.6 provide this. For a wedge the procedure given by Lewin (1969) or Felsen and Marcuvitz (1973) involves replacing the Bessel and Hankel functions in eqn. 7.53 by their integral forms and appropriate changes of the integration contour, but evaluation in known functions is apparently not possible without approximation.

Here an approximate solution found useful by Russo *et al.* (1965) is given. It is the asymptotic form of the far field of a line source near a wedge. It is derived by reciprocity from the first term of Pauli's (1938) asymptotic series for plane wave diffraction by a wedge. For wedge angles which are not large this results in Fresnel integrals.

In Fig. 7.7 the far field at r, θ of the TE-polarised line source at r_0, θ_0 is

$$E_z^i \simeq \frac{\exp\{-jk[r - r_0 \cos(\theta - \theta_0)]\}}{\sqrt{kr}} \qquad (7.56)$$

There is a reflected far-field in $0 < \theta < \pi - \theta_0$

$$E_z^{refl.} \simeq -\frac{\exp\{-jk[r - r_0 \cos(\theta + \theta_0)]\}}{\sqrt{kr}} \qquad (7.57)$$

and everywhere a diffracted far-field

$$E_z^{diff.} \simeq \frac{\exp(-jkr)}{\sqrt{kr}} \left[v(r_0, \theta - \theta_0) - v(r_0, \theta + \theta_0) \right] \tag{7.58}$$

where

$$v(r_0, \alpha) = \frac{2}{n} \frac{\sin \pi/n \, |\cos \alpha/2|}{(\cos \pi/n - \cos \alpha/n)} \frac{\exp(jkr_0 \cos \alpha + j\pi/4)}{\sqrt{\pi}}$$

$$\times F[\sqrt{2kr_0} \, |\cos \alpha/2|] \tag{7.59}$$

For $n = 2$, the wedge is a half-plane and eqns. 7.56, 7.57 and 7.58 together give the exact far-field of a line source near a conducting half-plane.

For TM-polarisation, H_z replaces E_z in the above expressions, the $(-)$ sign in the non-exponential part of the reflected wave term becomes $(+)$ and the diffracted field is

$$H_z^{diff.} \simeq \frac{\exp(-jkr)}{\sqrt{kr}} \left\{ v(r_0, \theta - \theta_0) + v(r_0, \theta + \theta_0) \right\} \tag{7.60}$$

This brief selection of diffraction solutions for conductors with edges is sufficient for the purposes of the following chapters. For a thorough coverage and bibliography the reader is referred to the excellent book by Bowman *et al.* (1969).

Chapter 8

Geometrical theory of diffraction by edges

8.1 The far field in half-plane diffraction

The geometrical theory of diffraction is Keller's (1953) name for an approximate method of solving diffraction problems which combines the principles of geometrical optics with asymptotic diffraction theory. The method is an improvement over the Kirchhoff theory in dealing with aperture antennas in that it admits edge diffraction and interaction and allows fields in and behind the aperture plane to be calculated. Although not rigorous, it provides approximate solutions to problems for which rigorous solutions are unavailable. It also gives the correct asymptotic form in situations where there is an exact solution. For diffraction by plane conducting screens we begin with the asymptotic solution for diffraction by a conducting half-plane.

Asymptotic approximation to the Fresnel integral for large values of the argument can be obtained by integration by parts. If $a \gg 1$, eqn. 7.26 can be written

$$
\begin{aligned}
F(a) &= \int_a^\infty \frac{d[\exp(-j\tau^2)]}{-2j\tau} \\
&= \frac{\exp(-ja^2)}{2ja} - \frac{1}{4} \int_a^\infty \frac{d[\exp(-j\tau^2)]}{\tau^3} \\
&\simeq \frac{\exp(-ja^2)}{2ja} + 0(a^{-3})
\end{aligned}
\tag{8.1}
$$

Repeated integration by parts yields an asymptotic series for $F(a)$, but only its asymptotic form or the first term is required here. If $a \ll -1$, eqn. 7.27 is used and

$$
F(a) \simeq \sqrt{\pi} \exp(-j\pi/4) + \frac{\exp(-ja^2)}{2ja}
\tag{8.2}
$$

When these results are applied to the solution for TE-polarised plane wave by a conducting half-plane, eqn. 7.25, one obtains, for $|a| \gg 1$ and an angle of incidence $0 < \theta_0 < \pi$,

(1) $E_z \simeq \exp[\,jkr\cos(\theta - \theta_0)] - \exp[\,jkr\cos(\theta + \theta_0)]$

$$+ D(\theta_0,\theta)\,\frac{\exp(-jkr)}{\sqrt{r}}, \qquad 0 < \theta < \pi - \theta_0$$

(2) $E_z \simeq \exp[\,jkr\cos(\theta - \theta_0)] + D(\theta_0,\theta)\,\dfrac{\exp(-jkr)}{\sqrt{r}}, \qquad \pi - \theta_0 < \theta < \pi + \theta_0$

(3) $E_z \simeq D(\theta_0,\theta)\,\dfrac{\exp(-jkr)}{\sqrt{r}} \qquad\qquad\qquad \pi + \theta_0 < \theta < 2\pi$

where

$$D(\theta_0,\theta) = -\frac{\exp(-j\pi/4)}{2\sqrt{2\pi k}}\left\{\sec\left(\frac{\theta - \theta_0}{2}\right) - \sec\left(\frac{\theta + \theta_0}{2}\right)\right\} \qquad (8.3)$$

The geometrical optics field consists of an incident and reflected wave in 1, the region of reflection, an incident wave in the illuminated region 2, and is non-existent in the shadow region 3. In all regions, there is a diffracted field which, in the far field, appears to originate from the edge.

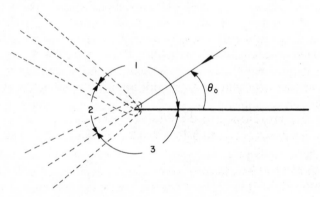

Fig. 8.1   *Geometrical optics regions in half-plane diffraction*
Within the parabolas, the asympotic expressions of eqn. 8.3 fail

On the boundaries $\theta = \pi \pm \theta_0$, the geometrical optics field is discontinuous and the coefficient of the diffracted field $D(\theta_0,\theta)$ is singular. The Fresnel integral arguments $a = \sqrt{2kr}\,\cos[(\theta \pm \theta_0)/2]$ are of equal magnitude M along curves defined by

$$r = \frac{M^2}{2k}\sec^2\left(\frac{\theta \pm \theta_0}{2}\right) \qquad (8.4)$$

which are parabolas with foci at the edge and axes along these boundaries, as shown in Fig. 8.1. If $M = 1$, for example, the parabolas have a semilatus sectum $r = \lambda/2\pi$ and the above expressions fail within the parabolas.

Returning to eqn. 7.25 and using

$$F(0) = \int_0^\infty \exp(-j\tau^2)\,d\tau = \frac{\sqrt{\pi}\,\exp(-j\pi/4)}{2}$$

at $\theta = \pi - \theta_0$

$$E_z \simeq \exp(-jkr\cos 2\theta_0) - \frac{\exp(-jkr)}{2} - \frac{\exp[-j(kr + \pi/4)]}{2\sqrt{2\pi kr}\,\sin\theta_0} \tag{8.5}$$

and at $\theta = \pi + \theta_0$

$$E_z \simeq \frac{\exp(-jkr)}{2} + \frac{\exp[-j(kr + \pi/4)]}{2\sqrt{2\pi kr}\,\sin 2\theta_0} \tag{8.6}$$

Hence, near the reflection and shadow boundaries the diffraction field is of the same order as the incident field. As $kr \to \infty$, transition between the geometrical optics fields in adjacent sectors is by their arithmetic mean. The boundary region singularities present one of the difficulties in applying the geometrical theory of diffraction.

The asymptotic solution for a TM-polarised plane wave is similar, with a (+) sign before the reflected field in region 1 and the second term in eqn. 8.3.

8.2 Keller's geometrical theory of diffraction for edges

In the geometrical optics field of the above half-plane solution, incident and reflected rays travel in straight lines and the reflected ray makes the same angle with the reflecting plane as the incident ray which produced it. Similarly, the diffracted field may be considered as an infinite system of rays travelling in straight lines outwards from the edge and originating from the incident ray which struck the edge. They are diffracted in a direction which makes the same angle with the edge as the incident ray and are quantitatively related to the field on the incident ray at the point of diffraction. Thus, when a ray is normally incident upon a straight edge, the field on a diffracted ray is

$$A^i D(\theta_0, \theta)\,\frac{\exp(-jkr)}{\sqrt{r}} \tag{8.7}$$

at r, θ with respect to the edge. A^i is the complex amplitude of the field on the incident ray at the point of diffraction and $D(\theta_0, \theta)$ is called the diffraction coefficient. It may also be assumed that any diffracted ray striking an edge can produce diffracted rays so that multiply diffracted fields can be produced. The field at any point will be the sum of the fields on all rays, incident, reflected, and diffracted, which pass through that point.

The assumption that diffraction is a local effect and the introduction of diffracted rays are the essential features of Keller's geometrical theory of diffraction. This theory has been applied to a variety of diffraction problems in electromagnetic theory, acoustics and elasticity. Wherever comparisons are possible, there is agreement with the asymptotic form of the exact solution. Results obtained by this method are valid when the diffracting object is large in wavelengths but good agreement has also been found for wavelengths as large as the diffracting object.

The problem of diffraction by a slit (Keller, 1957, 1962) serves to illustrate the method.

8.3 Diffraction by a slit

8.3.1 Single diffraction

A TE-polarised plane wave

$$E_z^i = \exp[jk(x \cos \theta_0 + y \sin \theta_0)] \tag{8.8}$$

is incident from $y > 0$ on the slit $|x| < a$ in a conducting plane in $y = 0$, as shown in Fig. 8.2. Diffracted rays are excited at the two edges $x = \pm a$ and their amplitude

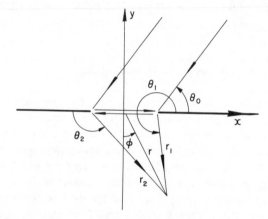

Fig. 8.2   *Diffraction by a slit*

and phase are given by eqn. 8.7. The singly diffracted field is the sum of the fields of these diffracted rays.

$$
\begin{aligned}
E_z^{sing.\,diff.} ={} & \exp(jka \cos \theta_0)D(\theta_0,\theta_1)\frac{\exp(-jkr_1)}{\sqrt{r_1}} \\
& + \exp(-jka \cos \theta_0)D(\pi + \theta_0,\theta_2)\frac{\exp(-jkr_2)}{\sqrt{r_2}}
\end{aligned} \tag{8.9}
$$

and the coordinates r_1,θ_1, refer to the edge at $x = a$ and r_2,θ_2 to the edge at $x = -a$.

The far-field pattern of the slit is obtained by letting

$$r_1 = r - a \sin \phi, \qquad \theta_1 = \frac{3\pi}{2} + \phi$$

$$\tag{8.10}$$

$$r_2 = r + a \sin \phi, \qquad \theta_2 = \frac{\pi}{2} + \phi$$

Then,

$$E_z^{sing.\ diff.} = \frac{\exp[-j(kr + \pi/4)]}{\sqrt{r\lambda}} f_s(\phi) \tag{8.11}$$

where

$$f_s(\phi) = \frac{j \sin[ka(\sin \phi + \cos \theta_0)]}{k \sin[\frac{1}{2}(\phi - \theta_0 + \pi/2)]} - \frac{\cos[ka(\sin \phi + \cos \theta_0)]}{k \cos[\frac{1}{2}(\phi + \theta_0 - \pi/2)]} \tag{8.12}$$

is the diffraction pattern of the singly diffracted fields.

On the shadow boundaries of the two half-planes $\theta_1 = \pi + \theta_0, \theta_2 = \theta_0, D(\theta_0, \theta)$ is singular and the field from each edge is not defined by eqn. 8.9. In the far field, however, the singularities on the shadow boundaries cancel and produce the effect of the incident field. Thus, neglecting interaction between the edges, the complete field in $y < 0$ is given by eqn. 8.11. Since the field on a diffracted ray diminishes as $(kr)^{-1/2}$, for wide apertures (small λs), the contribution from edge interaction is small and eqn. 8.12 satisfactorily represents the diffraction pattern of the slit, as shown in Fig. 8.2. For normal incidence on the slit and ϕ not large, essentially this result is obtained from Kirchhoff diffraction theory, taking as the aperture field the field of the incident wave in the slit.

8.3.2 Multiple diffraction

The singly diffracted ray from the right-hand edge of the slit in the direction $\theta_1 = \pi$ strikes the left hand edge in Fig. 8.2 when $r_1 = 2a$, and is diffracted again. So also is the singly diffracted field from the left hand edge in the direction $\theta_2 = \pi$ and the doubly diffracted field is

$$E_z^{double\ diff.} = \exp(jka \cos \theta_0) D(\theta_0, 2a, \pi) D(\pi, r_2, \theta_2)$$

$$+ \exp(-jka \cos \theta_0) D(\pi + \theta_0, 2a, \pi) D(\pi, r_1, \theta_1) \tag{8.13}$$

where

$$D(\theta_0, r, \theta) = D(\theta_0, \theta) \frac{\exp(-jkr)}{\sqrt{r}} \tag{8.14}$$

Doubly diffracted rays in the directions $\theta_1 = \theta_2 = \pi$ produce triply diffracted fields, and so on. Two multiply diffracted rays of each order will contribute to the field behind the screen. These can be written

$$E_z^{multiple\ diff.} = \exp(jka \cos \theta_0) D(\theta_0, 2a, \pi)[D(\pi, r_2, \theta_2)$$

$$+ D(\pi, 2a, \pi) D(\pi, r_1, \theta_1) + D^2(\pi, 2a, \pi) D(\pi, r_2, \theta_2) + \ldots]$$

$$+ \exp(-jka \cos \theta_0)D(\pi + \theta_0, 2a, \pi)[D(\pi, r_1, \theta_1)$$

$$+ D(\pi, 2a, \pi)D(\pi, r_2, \theta_2) + D^2(\pi, 2a, \pi)D(\pi, r_1, \theta_1) + \dots]$$

$$(8.15)$$

Adding the multiply diffracted field contributions results in a geometric series which can be summed

$$\sum_{n=1}^{\infty} D^{2n-2}(\pi, 2a, \pi) = \frac{1}{1 - D^2(\pi, 2a, \pi)} \tag{8.16}$$

Far from the slit, eqns. 8.10 may be used and the total diffracted field written

$$E_z = \frac{\exp[-j(kr + \pi/4)]}{\sqrt{\lambda r}} [f_s(\phi) + f_m(\phi)] \tag{8.17}$$

where $f_s(\phi)$ is defined by eqn. 8.12 and the pattern of the multiply diffracted fields is, for normal incidence ($\theta_0 = \pi/2$),

$$f_m(\phi) = \frac{4}{k} \{1 + \sqrt{4\pi ka} \exp[j(2ka + \pi/4)]\}^{-1}$$

$$\times \left(\frac{\cos(\phi/2) \cos[ka \sin \phi] + j \sin(\phi/2) \sin[ka \sin \phi]}{\cos \phi} \right) \tag{8.18}$$

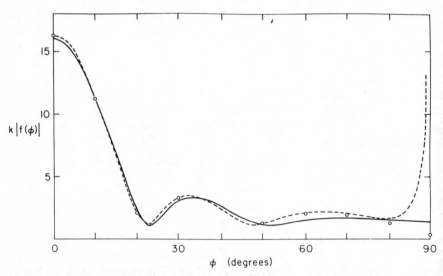

Fig. 8.3 *Diffraction pattern of a slit of width 2a = 2·55λ*
—— single diffraction eqn. 8.12
-- - - single & multiple diffraction eqns. 8.12 and 8.18
0 0 exact values (Karp and Russek, 1955)
                    (After Keller, 1957)

The diffraction pattern $|k(f_s(\phi) + f_m(\phi))|$ is shown in Fig. 8.3. Notice that the multiply diffracted fields are singular for $\phi = \pm \pi/2$, where there are shadow boundaries of the multiply diffracted fields.

The transmission cross-section σ of an aperture is defined as the ratio of the power transmitted through the aperture to the power incident on the aperture. It can be simply obtained from the imaginary part of the diffraction pattern of an aperture in the foreward direction (see Appendix A.6). Because of shadow boundaries at $\phi = \pm\pi/2$, the transmission cross-section of a slit, given by the imaginary part of $f_s(\theta_0 - \pi/2) + f_m(\theta_0 - \pi/2)$ is most accurate when obtained for normal incidence $(\theta_0 = \pi/2)$ by this method. This value is

$$\frac{\sigma}{2a} = 1 - \frac{\cos(2ka - \pi/4)}{\sqrt{\pi}\,(ka)^{3/2}}\left[1 - \frac{\sin(2ka - \pi/4)}{\sqrt{\pi ka}} + \frac{1}{4\pi ka}\right]^{-1} \tag{8.19}$$

In eqn. 8.19, the first term obtained from the imaginary part of $f_s(\theta_0 - \pi/2)$, as given by eqn. 8.13, is the geometrical optics cross-section. The second term, without the bracketed [ ] term, represents double diffraction and the bracketed [ ] higher-order multiple diffraction. It is apparent from the solid curve in Fig. 8.4*a*

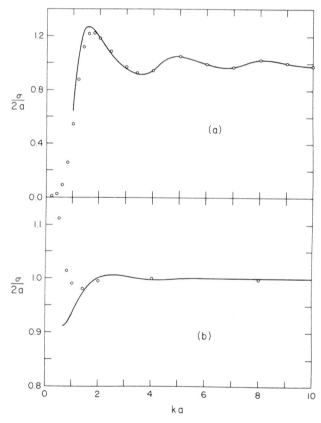

Fig. 8.4 *Transmission cross-section of a slit of width 2a for normal incidence of a plane wave.*
a TE polarisation, eqn. 8.19
b TM polarisation, eqn. 8.25
o o Exact values (Skavlem, 1952)
(After Keller, 1957)

that single and double diffraction alone provide a good approximation to the transmission cross-section for slits with $ka > 2$. The reason that higher order diffraction by the geometrical theory of diffraction evidently does not improve much the result for smaller ka is in part due to the omission of terms proportional to $(kr)^{-3/2}$, $(kr)^{-5/2}$, . . . from the asymptotic solution for half-plane diffraction (see eqn. 8.1). These higher-order diffracted fields become significant for small slit widths. Another reason is that a plane wave diffraction coefficient was used whereas the edges resemble line sources. In the 'self-consistent method' of Karp and Russek (1956) this assumption is not used and there is some improvement but their method is not as generally applicable as Keller's (1957). A comparison of results obtained with cylindrical wave diffraction by a strip was given by Yu and Rudduck (1967).

Interaction terms omitted from eqn. 8.19 are $O[(ka)^{-5/2}]$. If these terms and terms $O[(ka)^{-3}]$ are included by another high-frequency solution using equivalent line currents on the edges (Miller, 1958) a modest improvement in agreement with exact results can be obtained for $ka > 1 \cdot 4$. Clearly in this range of aperture widths it is simpler to use instead a series expansion of the field in powers of ka; indeed this must be done when $ka < 1$, where the asymptotic results are inaccurate. The two approaches overlap over a range of values of ka. In Fig. 8.4 the values for small ka are from an exact series solution by Skavlem (1951).

8.3.3 TM polarisation

For a TM polarised plane wave incident on the slit the singly diffracted field is given by eqn. 8.9 with H_z replacing E_z and a diffraction coefficient

$$D(\theta_0,\theta) = -\frac{\exp(-j\pi/4)}{2\sqrt{2\pi k}}\left[\sec\frac{(\theta-\theta_0)}{2} + \sec\frac{(\theta+\theta_0)}{2}\right] \tag{8.20}$$

This field vanishes in the aperture ($\theta = \pi$). Hence, in accordance with eqn. 7.3, if only the tangential magnetic field is considered it might be concluded incorrectly there is not interaction between the aperture edges. However, the tangential electric field $E_x = (j\omega\epsilon_0)^{-1}\partial H_z/\partial y$ does not vanish in the aperture. For interaction across the aperture, therefore, the diffracted field is proportional to the normal derivative $\partial A^i/\partial n$ of the field incident upon the diffracting edge. Rather than eqn. 8.7, the field on a diffracted ray is

$$\frac{\partial A^i}{\partial n}D'(\theta)\frac{\exp(-jkr)}{\sqrt{r}} \tag{8.21}$$

where $D'(\theta)$ is a new diffraction coefficient derived by Karp and Keller (1961). Here

$$D'(\theta) = \frac{1}{jk}\frac{\partial}{\partial\theta_0}D(\theta_0,\theta)\bigg|_{\theta_0=\pi} = \frac{-\exp(j\pi/4)}{2\sqrt{2\pi}k^{3/2}}\frac{\cos(\theta/2)}{\sin^2(\theta/2)} \tag{8.22}$$

To calculate the field resulting from TM double diffraction of the field singly diffracted at $(a,0)$, the normal derivative of the field on this ray at $(-a,0)$ is required. This is

$$\frac{\partial A^i}{\partial n} = -\exp(jka\cos\theta_0)\frac{1}{2a}\frac{\partial}{\partial\theta_1}\left[D(\theta_0,\theta_1)\frac{\exp(-jkr)}{\sqrt{r_1}}\right]_{\substack{r_1=2a\\ \theta_1=\pi}}$$

$$= \frac{\exp[-jka(2-\cos\theta_0)-j\pi/4]}{8(\pi ka^3)^{1/2}}\frac{\cos(\theta_0/2)}{\sin^2(\theta_0/2)} \tag{8.23}$$

and the doubly diffracted field from the left edge of Fig. 8.2 is

$$H_z^{double\ diff.} = -\frac{\exp[-jka(2-\cos\theta_0)-jkr_2]}{16\pi(ka)^{3/2}(2kr_2)^{1/2}}$$

$$\times\ \frac{\cos(\theta_0/2)\cos(\theta_2/2)}{\sin^2(\theta_0/2)\sin^2(\theta_2/2)} \tag{8.24}$$

The other multiply diffracted field contributions are calculated similarly (Karp and Keller, 1961). These aperture edge interaction terms are notably weaker than the corresponding terms for TE polarisation. This is apparent in the transmission cross-section, which for normal incidence is

$$\frac{\sigma}{2a} = 1 - \frac{\sin(2ka-\pi/4)}{8\sqrt{\pi}(ka)^{5/2}}\left[1 + \frac{\cos(2ka-\pi/4)}{8\sqrt{\pi}(ka)^{3/2}} + \frac{1}{256\pi(ka)^3}\right]^{-1} \tag{8.25}$$

in which omitted higher order terms are $0[(ka)^{-7/2}]$. The form of eqn. 8.25 (e.g. Jull, 1968a) differs in the bracketed term [ ] from that given by Karp and Keller (1961), but is more suitable for comparison with eqn. 8.19. This bracketed term, which includes triple and higher-order diffraction, is negligible for $ka > 1\cdot5$, and for $ka < 1\cdot5$ eqn. 8.25 fails (see Fig. 8.4b). As with TE polarisation, multiple diffraction beyond double does not significantly improve results. Values for small ka are again from the exact series solution of Skavlem (1951).

8.3.4 Diffraction by a strip

Since a slit in a conducting screen is the Babinet complement of a strip, solutions for high-frequency scattering by a strip are given by expressions in Sections 8.3.1–3. Thus, eqn. 8.17 with H_z replacing E_z gives the geometrical theory of diffraction solution for TM-polarised plane wave incidence on a conducting strip in $y = 0$, $|x| < a$. The scattering cross-section of a strip is the ratio of the power scattered by the strip per unit length in the z-direction to the power incident upon the strip. For a TM-polarised plane wave normally incident upon a strip of width $2a$ the scattering cross-section is given by twice eqn. 8.19; twice because power is scattered in the forward and rear directions. Similarly, the scattering cross-section of a normally incident TE-polarised plane wave is given by twice eqn. 8.25.

8.4 Diffraction by a curved edge

The field on an edge diffracted ray is given by eqn. 8.7 only in the special case when the diffracted wave is cylindrical, i.e. for normal incidence on a straight edge. A more general expression for the field on a diffracted ray is

$$A^i D(\theta_0, \theta) \frac{\exp(-jkr)}{\sqrt{r(1 + \rho_1^{-1} r)}} \tag{8.26}$$

where ρ_1 is the distance along the ray from the edge to a caustic, or confluence of the diffracted rays, as in Fig. 8.5. This result is obtained by considering the field amplitudes A and A_0 on two small portions dS and dS_0 of parallel wave fronts separated by a distance r (Fig. 8.6). Conservation of energy flux requires $A^2 dS = A_0^2 dS_0$ and since the areas of the wavefronts are proportional to the product of their radii of curvature

$$\frac{A}{A_0} = \left(\frac{dS_0}{dS}\right)^{1/2} = \left[\frac{\rho_1 \rho_2}{(\rho_1 + r)(\rho_2 + r)}\right]^{1/2}$$

or

$$A\rho_1^{-1/2} \left[(\rho_1 + r)(\rho_2 + r)\right]^{1/2} = A_0 \rho_2^{1/2} \tag{8.27}$$

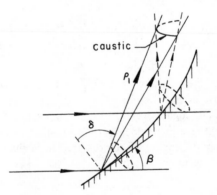

Fig. 8.5 *Diffraction by a curved edge*
Parallel rays incident on the edge produce cones of diffracted rays which intersect at a caustic ρ, along a diffracted ray from the edge.

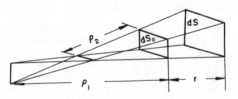

Fig. 8.6 *Portions of parallel wave fronts dS_0 and dS.*
The radii of curvature of dS_0 are ρ_1 and ρ_2 and of dS, $\rho_1 + r$ and $\rho_2 + r$

In eqn. 8.26, r is the distance along a ray from the edge, a caustic of the diffracted rays, so in eqn. 8.27 let $\rho_2 \to 0$. Since A is constant, the left-hand side of eqn. 8.27 must become a constant, say A_0', as $\rho_2 \to 0$. Then along a diffracted ray the field amplitude is

$$A = \frac{A_0'}{[r(1 + \rho_1^{-1} r)]^{1/2}}$$

which accounts for the r behaviour in eqn. (8.26).

The distance ρ_1 from the edge to the other caustic can be determined from the geometry (Keller, 1957, p. 442).

$$\rho_1 = -\frac{\rho \sin^2 \beta_0}{\rho \dot{\beta}_0 \sin \beta_0 + \cos \delta} \tag{8.28}$$

where ρ is the radius of curvature of the edge, β_0 is the angle between the incident ray direction and the tangent to the edge, $\dot{\beta}_0 = \partial \beta_0 / \partial s$, where s is the arc length along the edge, and δ is the angle between the normal to the edge and the diffracted ray.

8.3 Diffraction by a circular aperture

An example to the preceding Section, consider a TE-polarised plane wave normally incident upon a circular aperture of radius a. Then $\dot{\beta}_0 = 0$ and for normal incidence $\beta_0 = \pi/2$, so $\rho_1 = -\rho/\cos \delta = a/\cos \theta$, since $\rho = a$ and $\delta = \pi - \theta$. The incident wave is $E_z^i = \exp(jky)$ as in Fig. 8.2 with $\theta_0 = \pi/2$. Singly diffracted rays from the nearest and farthest points on the aperture edge pass through each point not on the axis, in accordance with the extension of Fermat's principle to diffracted rays. The field of the singly diffracted rays is

$$E_z^{sing.\ diff.} = \frac{\exp(-jkr_1)}{\sqrt{r_1 (1 - (r_1/a) \cos \theta_1)}} D\left(\frac{\pi}{2}, \theta_1\right)$$
$$+ \frac{\exp(-jkr_2)}{\sqrt{r_2 (1 - (r_2/a) \cos \theta_1)}} D\left(\frac{3\pi}{2}, \theta_2\right) \tag{8.29}$$

where $D(\theta_0, \theta)$ is given by eqn. 8.3. Using the far-field expressions of eqn. 8.10,

$$E_z^{single\ diff.} = \frac{\exp(-jkr)}{\lambda r} f_s(\phi) \tag{8.30}$$

where

$$f_s(\phi) = \frac{1}{k^{3/2}} \sqrt{\frac{2\pi a}{\sin \phi}} \left[j \frac{\sin(ka \sin \phi - \pi/4)}{\sin(\phi/2)} - \frac{\cos(ka \sin \phi - \pi/4)}{\cos(\phi/2)} \right] \tag{8.31}$$

is the diffraction pattern for single diffraction. This result is singular on the axis ($\phi = 0$), which is both a caustic of the singly diffracted rays and a shadow boundary. Keller (1957) used the exact solution to correct for this, or

$$f_s(\phi) = \frac{\pi a}{k} \left[\frac{jJ_1(ka \sin \phi)}{\sin(\phi/2)} - \frac{J_0(ka \sin \phi)}{\cos(\phi/2)} \right] \tag{8.32}$$

for $ka \sin \phi$ less than a constant for a given accuracy. For example, if $ka = 3\pi$, eqn. 8.32 is required for $\phi < 0{\cdot}1$ radians (Keller, 1957, Fig. 10).

Multiply diffracted fields can be obtained as in Section 8.2.2. If

$$D(\theta_0, r, \theta) = \frac{\exp(-jkr)}{\sqrt{r(1 + (r/a) \cos \theta)}} D(\theta_0, \theta) \tag{8.33}$$

the multiply diffracted field is given by eqn. 8.15 with $\theta_0 = \pi/2$. Since $D(\pi/2, 2a, \pi) = D(3\pi/2, 2a, \pi)$ this reduces to

$$E_z^{mult. \ diff.} = \frac{D(\pi/2, 2a, \pi)[D(\pi, r_1, \theta_1) + D(\pi, r_2, \theta_2)]}{1 - D^2(\pi, 2a, \pi)} \tag{8.34}$$

and the far field is obtained by using eqn. 8.10. The axis $\phi = 0$ is a caustic of all multiply diffracted fields and, as with the singly diffracted fields, a correction based on the exact solution is required near $\phi = 0$. With this correction the transmission cross-section for normal incidence is

$$\frac{\sigma}{\pi a^2} = 1 - \frac{2 \sin(2ka - \pi/4)}{\sqrt{\pi}(ka)^{3/2}} \left[1 + \frac{\cos(2ka - \pi/4)}{\sqrt{\pi ka}} + \frac{1}{4\pi ka} \right]^{-1} \tag{8.35}$$

Numerical values for this transmission cross-section (Keller, 1957, Fig. 11) show that, as with the slit, omission of the bracketed term [ ] leaving only single and double diffraction provides a good approximation to exact values for $ka > 2$.

For TM polarisation, H_z replaces E_z and the singly diffracted field is given by eqn. 8.31 or 8.32 with the $(-)$ sign between the two terms replaced by $(+)$. Double diffraction is calculated by using eqn. 8.21 with a factor $(1 + \rho_1^{-1} r)^{-1/2}$ on the right-hand side. Again these fields are singular on the axis and must be corrected there. With this correction the transmission cross-section for single and double diffraction is

$$\frac{\sigma}{\pi a^2} = 1 - \frac{\sin(2ka + \pi/4)}{16\sqrt{\pi}(ka)^{7/2}} \tag{8.36}$$

which is accurate for $ka > 2$ (Keller, 1962, Fig. 12).

8.6 Diffraction coefficients for oblique incidence and for a wedge

For a TE-polarised plane wave obliquely incident on a half plane, the diffracted field can be obtained from the asymptotic form of eqn. 7.39. Using eqn. 8.1 this is

$$E_z^{diff.} \simeq - \frac{\sin \beta_0 \exp[jk(z \cos \beta_0 - r \sin \beta_0) - j\pi/4]}{2\sqrt{2\pi k \sin \beta_0 r}}$$

$$\times \left\{ \sec\left(\frac{\theta - \theta_0}{2}\right) - \sec\left(\frac{\theta + \theta_0}{2}\right) \right\} \tag{8.37}$$

where r is the radial distance from the edge. If the distance along a ray from the diffraction point on the edge is $s = r/\sin \beta_0 = -z/\cos \beta_0$, eqn. 8.37 can be written

$$E_z^{diff.} \simeq \sin \beta_0 \, \frac{\exp(-jks)}{\sqrt{s}} \, D(\theta_0, \theta) \tag{8.38}$$

with a diffraction coefficient

$$D(\theta_0, \theta) = -\frac{\exp(-j\pi/4)}{2\sqrt{2\pi k} \sin \beta_0} \left\{ \sec \frac{(\theta - \theta_0)}{2} - \sec \frac{(\theta + \theta_0)}{2} \right\} \tag{8.39}$$

For TM polarisation, the diffraction coefficient is as in eqn. 8.39 with the $(-)$ sign between the terms replaced by a $(+)$.

The asymptotic form of the exact solution for plane wave diffraction by the perfectly conducting wedge of angle $(2-n)\pi$ in Fig. 7.1 reveals a diffracted field of the form of eqn. 8.38 with a diffraction coefficient

$$D(\theta_0, \theta) = \frac{\exp[-j(\pi/4)]}{\sqrt{2\pi k} \sin \beta_0} \frac{\sin(\pi/n)}{n}$$

$$\times \left\{ \frac{1}{\cos(\pi/n) - \cos[(\theta - \theta_0)/n]} \mp \frac{1}{\cos(\pi/n) - \cos[(\theta + \theta_0)/n]} \right\} \tag{8.40}$$

where the upper sign applies for TE polarisation and the lower sign for TM polarisation. This diffraction coefficient reduces to the half-plane form when $n = 2$. These diffraction coefficients all fail on shadow boundaries.

8.7 Uniform solutions

8.7.1 Uniform diffraction theories

Keller's geometrical theory of diffraction fails on and near caustics and shadow boundaries. As all ray methods experience difficulties at caustics there other methods are recommended. Ufimtsev's (1962, 1975) method of edge waves is applicable as is the related method of equivalent edge currents used by Millar (1956, 1957) for plane screens and Ryan and Peters (1969) for wedges. Both techniques have been compared with the geometrical theory of diffraction by Knott and Senior (1974). Alternatively, the Kirchhoff method is usually applicable and accurate for axial caustics of apertures.

Several procedures have evolved to obtain the fields in the transition regions around shadow and reflection boundaries, such as in Section 8.1, using ray techniques and the concepts of the geometrical theory of diffraction. The simplest way to eliminate these boundary region singularities for two-dimensional fields is to use the solution for the far field of an isotropic line source near a half-plane, eqn. 7.51, and for a wedge the expressions in Section 7.7.2 may be adopted. These are used in Chapter 9. This informal approach has been used by Ohba (1961, 1963) for plane screens with straight edges, Rudduck (1965) for a straight wedge and James (1974) for a plane screen with a curved edge. The most complete generalisation of this kind of approach is by Kouyoumjian and Pathak (1974), who constructed a uniform geometrical theory of diffraction for curved edges and wedges with a dyadic diffraction coefficient applicable for plane, cylindrical or spherical wave incidence. For example, their diffracted field of the wedge of Fig. 7.7, uniformly valid at all points away from the edge, is given by eqn. 8.26 with s replacing r and a diffraction coefficient

$$D(\theta_0,\theta) = \frac{-\exp[-j(\pi/4)]}{2n\sqrt{2\pi k}\,\sin\beta_0}\left[\cot\left(\frac{\pi + (\theta - \theta_0)}{2n}\right) G[kLa^+(\theta - \theta_0)]\right.$$

$$+ \cot\left(\frac{\pi - (\theta - \theta_0)}{2n}\right) G[kLa^-(\theta - \theta_0)]$$

$$\mp \left\{\cot\left(\frac{\pi + (\theta + \theta_0)}{2n}\right) G[kLa^+(\theta + \theta_0)]\right.$$

$$\left.\left. + \cot\left(\frac{\pi - (\theta + \theta_0)}{2n}\right) G[kLa^-(\theta + \theta_0)]\right\}\right] \qquad (8.41)$$

with

$$G(x) = 2j\,|\sqrt{x}\,|\exp(jx)F(\,|\sqrt{x}\,|\,) \qquad (8.42)$$

and $F(x)$ the Fresnel integral given in eqn. 7.26. In eqn. 8.41,

$$a^\pm(\beta) = 2\cos^2\left(\frac{2n\pi N^\pm - \beta}{2}\right) \qquad (8.43)$$

in which $N^\pm$ are the integers which most nearly satisfy the equations

$$2\pi nN^+ - \beta = \pi$$
$$2\pi nN^- - \beta = -\pi \qquad (8.44)$$

Here $\beta = \theta \pm \theta_0$ and so N^+ and N^- each have two values. For an exterior edge $(1 < n \leqslant 2)$, $N^+ = 0$ or 1 and $N^- = -1$, 0 or 1. L is a distance parameter which for plane, cylindrical and spherical wave incidence is, respectively,

$$L = \begin{cases} s\sin^2\beta_0 \\[2mm] \dfrac{rr_0}{r + r_0} \\[2mm] \dfrac{ss_0}{s + s_0}\sin^2\beta_0 \end{cases} \qquad (8.45)$$

For $ks \gg 1$ and outside the transition regions, eqn. 8.41 becomes eqn. 8.40. At a shadow or reflection boundary a cotangent function in eqn. 8.41 becomes singular but its product with the transition function is bounded. Grazing incidence ($\theta_0 = 0, n\pi$) must be considered separately, but for TE polarisation a uniform solution follows from differentiating eqn. 8.41 as $D'(\theta)$ in eqn. 8.21 (Kouyoumjian, 1975).

The advantage of this method is its relative simplicity; its main disadvantage is that it is approximate for multiple diffraction or nonisotropic sources. In most practical applications the former consideration will outweigh the latter. The theory has subsequently been modified by the addition of slope diffraction to include the effect of the pattern of the source (Kouyoumjian and Pathak, 1977).

One way of obtaining the field on the shadow boundaries of edges on plane screens for arbitrary sources consists of integrating the product of the pattern function of the source and the solution for plane wave diffraction by a half-plane. This was Clemmow's (1950) approach mentioned in Section 7.6 for an isotropic line source parallel to the edge of a half-plane. Its extension to anisotropic sources is proposed by Mittra *et al.* (1976) and Rahmat-Samii and Mittra (1977) as a 'spectral theory of diffraction'. The field on caustics of, for example, the circular aperture can also be obtained in this way.

8.7.2 Uniform asymptotic theory of diffraction

A uniform asymptotic theoy of diffraction which provides the correct asymptotic solution for an arbitrary incident field on a half-plane has been developed by Lewis and Boersma (1969) and Ahluwalia, Lewis and Boersma (1968). It also corrects other defects of Keller's theory in providing the complete field up to the edge and includes a systematic procedure for determining higher-order terms in the diffracted field expansion.

A description of the uniform asymptotic theory for a cylindrical wave from a line source at r_0, θ_0 parallel to the edge of a conducting half-plane as in Fig. 7.6 has been given by Boersma and Lee (1977) and is summarised here. All fields are expanded asymptotically in inverse powers of the wavenumber k, which is assumed large. The incident TE-polarised wave of amplitude $z^i(R,\phi)$ is expanded asymptotically as

$$E_z^i \simeq \exp(-jkR)z^i(R,\phi)$$

$$= \exp(-jkR) \sum_{m=0}^{\infty} (-jk)^{-m} z_m^i(R,\phi) \qquad (8.46)$$

where the amplitude coefficients $z_m^i(R,\phi)$ may be derived by inserting eqn. 8.46 in $(\nabla^2 + k^2)E_z = 0$, (Lewis and Boersma, 1969).

This incident field results in a total field

$$E_z = U(r,\theta) - U(r,4\pi - \theta) \qquad (8.47)$$

in which the first term is associated with the incident field of the line source and the second term with the image of the line source in the half plane (see Fig. 7.6). Lewis and Boersma (1969) postulated the following uniform asymptotic expansion for $U(r,\theta)$:

$$U(r,\theta) \simeq \exp[-jk(r_0 + r)]\Bigg\{ [H(p) - \hat{H}(p)]z^i(R,\phi) + k^{-1/2}$$

$$\times \sum_{m=0}^{\infty} (-jk)^{-m} v_m(r,\theta) \Bigg\} \qquad (8.48)$$

In eqn. 8.48,

$$H(x) = \frac{\exp(-j\pi/4)}{\sqrt{\pi}} \exp(jx^2) \int_{-x}^{\infty} \exp(-j\tau^2)d\tau \tag{8.49}$$

$$\simeq \exp(jx^2)u(x) + \hat{H}(x) \quad x \to \pm\infty \tag{8.50}$$

in accordance with eqn. 8.2. Here, $u(x)$ is the unit step function; $u(x) = 1$ for $x > 0$ and $u(x) = 0$ for $x < 0$. The full asymptotic expansion of the complex Fresnel integral (see eqn. 8.1) gives

$$\hat{H}(x) = \frac{-\exp(-j\pi/4)}{2\pi x} \sum_{m=0}^{\infty} \Gamma(m + 1/2)(-jx^2)^{-m} \tag{8.51}$$

with the gamma function

$$\Gamma(m + 1/2) = \sqrt{\pi} \cdot \frac{1}{2} \cdot \frac{3}{2} \cdots \cdots \frac{2m - 1}{2}$$

The argument of the Fresnel integral is

$$p = 2 \sqrt{\frac{krr_0}{r + r_0 + R}} \cos\left(\frac{\theta - \theta_0}{2}\right) \tag{8.52}$$

The first two terms in the final asymptotic expansion of eqn. 8.48 are

$$v_0(r,\theta) = \frac{-\exp[-j(\pi/4)]}{2\sqrt{2\pi r}} z_0^i \sec\left(\frac{\theta - \theta_0}{2}\right) \tag{8.53}$$

$$v_1(r,\theta) = \frac{-\exp[-j(\pi/4)]}{2\sqrt{2\pi r}} \left\{ z_1^i \sec\left(\frac{\theta - \theta_0}{2}\right) \right.$$

$$+ \left[\frac{3}{2r_0} z_0^i \sin^2\left(\frac{\theta - \theta_0}{2}\right) - \frac{\partial z_0^i}{\partial R} \cos(\theta - \theta_0) + \right.$$

$$\left. + \frac{1}{r_0} \frac{\partial z_0^i}{\partial \phi} \sin(\theta - \theta_0) \right] \frac{1}{4} \sec^2\left(\frac{\theta - \theta_0}{2}\right)$$

$$\left. + \frac{1}{4r} z_0^i \sec^3\left(\frac{\theta - \theta_0}{2}\right) \right\} \tag{8.54}$$

where z_0^i and z_1^i and their derivatives are evaluated at the edge $R = r_0, \phi = \pi - \theta_0$. Higher-order v_m can be obtained from a recursive formula (Ahluwalia *et al.*, 1968) but are not usually needed.

For TM polarisation, H_z replaces E_z and the $(-)$ between the terms of eqn. 8.47 is replaced by $(+)$.

For plane wave incidence eqn. 8.47 reduces to the exact solution of eqn. 7.25. This is the basis for postulating eqn. 8.48. At far-field points not near shadow or reflection boundaries, that part of eqn. 8.48 associated with the incident wave amplitude $z^i(R,\phi)$ becomes the geometrical optics field and the remainder the diffracted field, fields which cannot be distinguished in the transition regions around these boundaries.

Fields calculated from the above expressions are uniformly valid for all $0 < r < \infty$, $0 \leqslant \theta \leqslant 2\pi$. Away from shadow and reflection boundaries with $|p| \gg 1$, the diffracted field of eqn. 8.3 is obtained, i.e. the result provided by Keller's geometrical theory of diffraction. On the shadow boundary $\theta = \pi + \theta_0, p = 0$ and both parts of U become infinite but their singularities cancel leaving a finite field. The function U remains finite everywhere and provides a smooth transition across shadow and reflection boundaries. Higher-order diffracted fields, such as that given by eqn. 8.21 are contained in the above expression.

This uniform asymptotic theory of diffraction can be regarded as a generalisation of Keller's geometrical theory of diffraction just as Keller's theory may be regarded as a generalisation of geometrical optics. Like the geometrical theory of diffraction, although not rigorously proven always to yield the correct asymptotic result, it has in all cases involving half-planes when put to the test. Unfortunately, these are comparatively few due to its complexity in applications where interaction between edges is significant (e.g. Boersma, 1975a, b, Lee and Boersma, 1975, Ciarkowski, 1975). The method has been extended to electromagnetic diffraction by a curved wedge by Lee and Deschamps (1976) but there it is approximate. Lee (1978) has reviewed uniform asymptotic edge diffraction.

Chapter 9

Applications of geometrical diffraction theory to antennas

9.1 Scope and limitations

In antenna analysis the geometrical theory of diffraction is particularly useful in obtaining the radiation field at angles well off the beam axis or in the rear direction where the Kirchhoff method either fails or cannot be applied. On or near the beam axis geometrical diffraction theory may fail, as in the case of a paraboloidal reflector. There the Kirchhoff method is both applicable and accurate and should be used. In this sense the two methods are complementary.

Shadow boundary singularities are usually encountered in any application of the geometrical theory of diffraction to antennas, for example in field calculations in the lateral directions of planar apertures. These difficulties can be avoided or overcome by the procedures of Section 8.7. Here the informal methods are usually used rather than the uniform asymptotic theory of Section 8.7.2.

It is important to remember this is a high-frequency technique which, like the Kirchoff method, diminishes in accuracy with the aperture dimensions in wavelengths. It is, however, surprisingly accurate for apertures which are not large. Figs. 8.4 indicate that aperture widths should be larger than about a half-wavelength. They also indicate that multiple diffraction beyond double provides little improvement with the geometrical theory of diffraction, because only the lowest order diffracted field is included in the diffraction coefficient. This attribute is generally shared with the methods of equivalent currents and edge waves (Ufimstsev, 1975). The next higher-order interaction can be included by slope diffraction (e.g. Rudduck and Wu, 1969) where it is needed. The uniform asymptotic theory of diffraction can be used to systematically include further higher order interaction but in practical applications the improvement obtained will rarely justify the effort required. Approximations are usually unavoidable in antenna analysis and tend to render refined analysis meaningless.

It is appropriate to begin with the analysis of an idealized antenna, the open-ended parallel plate waveguide. Exact results for the reflection coefficient and radiation pattern were obtained independently by Heins (1948) for TE polarisation and by Weinstein (1948, 1966) for both polarisations by solutions of integral equations of the Wiener-Hopf type (see also Collin and Zucker, 1969). These exact solutions provide a valuable test of ray methods in situations where shadow boundaries are present and edge interaction is significant. Some results from the uniform asymptotic theory may also be included for comparison. It is found that for this type of aperture ray methods can be accurate for aperture widths smaller than those indicated by results for the slit. These results are also of practical value in the analysis of antennas such as open-ended rectangular waveguides and sectoral horns.

9.2 Reflection from open-ended parallel-plate waveguides

9.2.1 TM polarisation

A ray solution for TEM mode reflection from the open end of a parallel-plate waveguide was obtained by Rudduck and Tsai (1968) who included only dominant mode single diffraction by the operative edges. Yee *et al.* (1968) included all incident modes and full interaction in their ray optical solution along with a useful procedure for converting ray fields to waveguide modes. Their method is followed here.

The waveguide in Fig. 9.1 is composed of conducting half-planes in $x > 0, y = 0$, a. A TM-polarised field incident from the waveguide is

$$H_z^i = \cos\left(\frac{N\pi y}{a}\right)\exp(jk_N x), \qquad N = 0,1,2 \ldots$$

$$= \frac{1}{2}\left\{\exp\left[j\left(k_N x + \frac{N\pi y}{a}\right)\right] + \exp\left[j\left(k_N x - \frac{N\pi y}{a}\right)\right]\right\} \qquad (9.1)$$

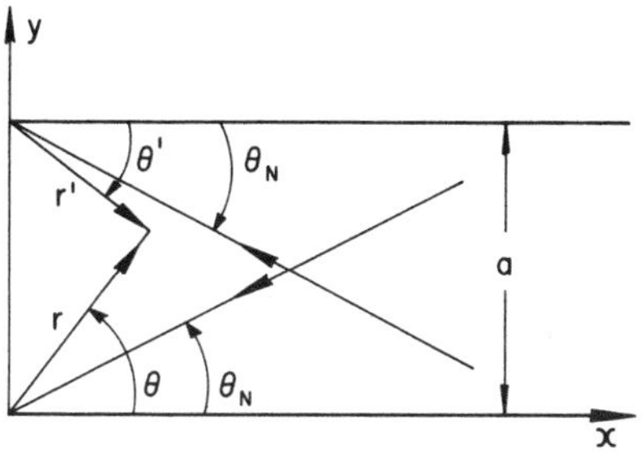

Fig. 9.1   *Coordinates for reflection from an open-ended parallel-plate waveguide*

where

$$k_N = [k^2 - (N\pi/a)^2]^{1/2}, \qquad k^2 > (N\pi/a)^2$$

The plane waves of eqn. 9.1 are incident on the lower and upper half-planes at angles θ_N and $2\pi - \theta_N$, respectively, where $\theta_N = \cos^{-1}(k_N/k)$. Incident field amplitudes at the edges are $1/2$ and $(-1)^N/2$. The diffracted fields are given by eqn. 8.7 with $D(\theta_0,\theta)$ defined by eqn. 8.20. Since $D(2\pi - \theta_N, 2\pi - \theta') = D(\theta_N,\theta')$ the singly diffracted (zero interaction) field of the edges are

$$H_z^{(0)} = \frac{1}{2} D(\theta_N,\theta) \frac{\exp(-jkr)}{\sqrt{r}} + \frac{(-1)^N}{2} D(\theta_N,\theta') \frac{\exp(-jkr')}{\sqrt{r'}} \qquad (9.2)$$

where the coordinates r, θ and r', θ' refer to the edges at $x = 0, y = 0$ and a, respectively. These fields appear to originate from line sources of the form

$$H_z = f(\theta) \frac{\exp[-j(kr - \pi/4)]}{\sqrt{r}} \qquad (9.3)$$

at the edges. According to Yee *et al.* (1968, eqn. (15)) the modal expansion of the field in the guide due to the line source of eqn 9.3 on the guide wall at $y_0 = 0$ or a is

$$H_z = \sum_{n=0}^{\infty} \frac{\epsilon_n \sqrt{2\pi k}}{2k_n a} f(\theta_n) \exp[-j(n\pi y_0/a)] \cos\left(\frac{n\pi y}{a}\right) \exp(-jk_n x) \qquad (9.4)$$

where $\epsilon_0 = 1$ and $\epsilon_n = 2, n \neq 0$ and

$$\left.\begin{array}{l} k_n = k \cos\theta_n = \sqrt{k^2 - (n\pi/a)^2} \\[2mm] \quad\quad\quad = -j\sqrt{(n\pi/a)^2 - k^2} \end{array}\right\} k^2 \gtrless \left(\frac{n\pi}{a}\right)^2 \qquad (9.5)$$

Eqn. 9.4 permits the ray field of eqn.9.2 to be rewritten as a mode sum. For the singly diffracted fields of the two edges

$$H_z^{(0)} = \sum_{n=0}^{\infty} \epsilon_n \Gamma_{Nn}^{(0)} \cos\left(\frac{n\pi y}{a}\right) \exp(-jk_n x) \qquad (9.6)$$

where $\Gamma_{Nn}^{(0)}$ is the contribution of the singly diffracted fields to the reflection coefficient of the TM_n mode for an incident TM_N mode

$$\Gamma_{Nn}^{(0)} = [1 + (-1)^{N+n}] \frac{\sqrt{2\pi k}}{4k_n a} D(\theta_N,\theta_n) \exp(-j\pi/4)$$

$$= \frac{j\sqrt{k + k_N} \sqrt{k + k_n}}{2k_n a(k_N + k_n)}, \qquad n + N \text{ even} \qquad (9.7)$$

and $\Gamma_{Nn}^{(0)} = 0$ for $N + n$ odd. There is no coupling between even and odd numbered waveguide modes. This holds for all orders of interaction between the edges and is a consequence of waveguide and mode symmetry about the $y = a/2$ plane.

The singly diffracted fields of eqn. 9.2 in the directions $\theta = \theta' = \pi/2$ are doubly diffracted by the opposite edges at $r = r' = a$. This first-order interaction between the aperture edges produces the doubly diffracted fields

$$H_z^{(1)} = \frac{1}{2} D\left(\theta_N, \frac{\pi}{2}\right) \frac{\exp(-jka)}{\sqrt{a}} \left\{ D\left(\frac{\pi}{2}, \theta'\right) \frac{\exp(-jkr')}{\sqrt{r'}} + (-1)^N D\left(\frac{\pi}{2}, \theta\right) \right.$$

$$\times \left. \frac{\exp(-jkr)}{\sqrt{r}} \right\}$$

$$= \sum_{n=0}^{\infty} \epsilon_n \Gamma_{Nn}^{(1)} \cos\left(\frac{n\pi y}{a}\right) \exp(-jk_n x) \tag{9.8}$$

where, according to eqn. 9.4, the first-order interaction contribution to the reflection coefficient is

$$\Gamma_{Nn}^{(1)} = \frac{\sqrt{2\pi k}}{4k_n a} D\left(\theta_N, \frac{\pi}{2}\right) \frac{\exp(-jka)}{\sqrt{a}} [(-1)^n + (-1)^N] D\left(\frac{\pi}{2}, \theta_n\right) \exp(-j\pi/4)$$

$$= (-1)^{N+1} \sqrt{\frac{ka}{2\pi} \frac{\sqrt{k+k_N} \sqrt{k+k_n}}{2k_n^2 k_N a^2}} \exp[-j(ka - \pi/4)] \tag{9.9}$$

for $N + n$ even and $\Gamma_{Nn}^{(1)} = 0$ for $N + n$ odd.

The doubly diffracted fields of eqn. 9.8 in the directions $\theta = \theta' = \pi/2$ are triply diffracted at $r = r' = a$ and so on, but the above procedure cannot be followed directly because $D(\pi/2, \pi/2)$ is singular. The aperture plane is a reflection boundary of the multiply diffracted fields. To include the higher-order interaction contributions to the reflection coefficient the edges excited on single diffraction are approximated by equivalent isotropic line sources and the expressions of Section 7.6 used.

To illustrate, the singly diffracted field from the lower edge in the direction of the upper edge is given by the first term of eqn. 9.2 with $\theta = \pi/2$.

$$\frac{\sqrt{k}}{2} D\left(\theta_N, \frac{\pi}{2}\right) \frac{\exp(-jkr)}{\sqrt{kr}} \tag{9.10}$$

This is an isotropic line source at $r_0 = a$, $\theta_0 = 3\pi/2$ from the upper half-plane. According to eqn. 7.45, with a (+) between the terms for TM polarisation, the asymptotic form of the field scattered from the upper edge on the reflection boundary $\theta' = \pi/2$ is

$$\frac{\sqrt{k}}{2} D\left(\theta_N, \frac{\pi}{2}\right) \left\{ \frac{1}{2} \frac{\exp[-jk(a+r')]}{\sqrt{k(a+r')}} + \frac{\exp(-j\pi/4)}{2} \frac{\exp[-jk(a+r')]}{\sqrt{2\pi ka} \sqrt{kr'}} \right\} \tag{9.11}$$

which contains both reflected and diffracted components. Putting $r' = a$ in eqn. 9.11 and multiplying by $r^{-1/2} \exp(-jkr) D(\pi/2, \theta)$ gives one contribution to the triply diffracted field. The other is from fields initially and finally diffracted at the upper edge. Together they constitute the second-order interaction fields of the edges,

$$H_z^{(2)} = D\left(\theta_N, \frac{\pi}{2}\right) \frac{\exp(-j2ka)}{4\sqrt{2a}} \left[1 + \frac{\exp(-j\pi/4)}{\sqrt{\pi ka}}\right] \left\{ D\left(\frac{\pi}{2}, \theta'\right) \frac{\exp(-jkr')}{\sqrt{r'}} \right.$$

$$\left. + (-1)^N D\left(\frac{\pi}{2}, \theta\right) \frac{\exp(-jkr)}{\sqrt{r}} \right\} \tag{9.12}$$

which can be converted to a modal expansion through eqn. 9.4. This procedure can be used to calculate further higher-order interactions between the half-planes. Adding the contribution from each multiply diffracted field traced individually results in a series. The composite diffracted field from each edge cannot be obtained by a self-consistent method such as was used by Karp and Russek (1956) for the slit because the scattering centers are not only the edges but also their images.

The total reflection coefficient into the nth mode due to single and multiple diffraction of an incident Nth mode is:

$$\Gamma_{Nn} = \Gamma_{Nn}^{(0)} \left\{ 1 + \frac{\exp(-j\pi/4)}{\sqrt{2\pi ka}} \frac{k(k_N + k_n)}{k_N k_n} A^{\pm}(ka) - \frac{j(k_N + k_n)}{4\pi ka k_N k_n} \right.$$

$$\left. \times \left[A^{\pm}(ka)\right]^2 + 0\left[(ka)^{-3/2}\right] \right\} \tag{9.13}$$

for $N + n$ even and $\Gamma_{Nn} = 0$ for $N + n$ odd. $\Gamma_{Nn}^{(0)}$ is given by eqn. 9.7 and in the series

$$A^{\pm}(ka) = \pm \sum_{m=1}^{\infty} \frac{(\mp 1)^{m-1} \exp(-jmka)}{2^{m-1} m^{1/2}} \tag{9.14}$$

the upper sign applies for N odd and the lower sign for N even. This series converges sufficiently rapidly for numerical calculations.

In Fig. 9.2, the amplitudes of the reflection coefficients of the lower order modes are compared with the exact results of Weinstein (1966). The latter are, for incident and reflected TEM modes ($N = n = 0$),

$$|\Gamma_{00}| = \exp(-\pi a/\lambda), \qquad\qquad 0 < a/\lambda < 1,$$

$$= \frac{(a/\lambda) + \gamma_0}{(a/\lambda) - \gamma_0} \exp(-\pi a/\lambda), \qquad 1 < a/\lambda < 2 \tag{9.15}$$

and for TEM-TM$_2$ reflection

$$|\Gamma_{02}| = \frac{2a}{\lambda} (2|\Gamma_{00}|)^{1/2} \exp(-\pi\gamma_0/2), \qquad 1 < a/\lambda < 2, \tag{9.16}$$

where $\gamma_0 = [(a/\lambda)^2 - 1]^{1/2}$. The phases of these exact reflection coefficients are more complicated.

From eqn. 9.7 with $k_0 = 2\pi/\lambda$ and $k_2 = [k^2 - (2\pi/a)^2]^{1/2}$ the noninteraction terms $\Gamma_{00}^{(0)}$ and $\Gamma_{02}^{(0)}$ give the monotonic broken curves in Fig. 9.2. The dashed curves include single interaction between the half-planes. For TEM-TEM reflection this is

$$| \Gamma_{00}^{(0)} + \Gamma_{00}^{(1)} | = \left| \frac{j}{2ka} - \frac{\exp[-j(ka - \pi/4)]}{\sqrt{2\pi}(ka)^{3/2}} \right| \qquad (9.17)$$

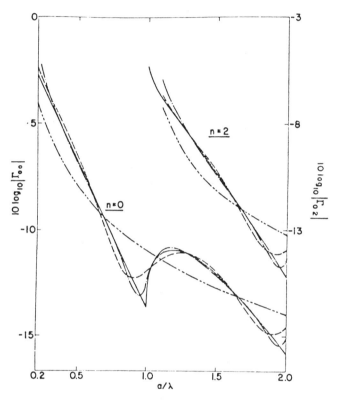

Fig. 9.2   *Reflection coefficients for TEM mode incidence on the open end of a parallel-plate waveguide of width a*
—–—   single diffraction $(|\Gamma^{(0)}|)$
------   single and double diffraction $(|\Gamma^{(0)} + \Gamma^{(1)}|)$
—·—   single and multiple diffraction eqn. 9.13
———   exact, eqns. 9.15 and 9.16

which is a good approximation to the exact result except near mode cutoff widths. Inclusion of higher-order interactions through eqn. 9.13 provides some improvement at the expense of a more complicated formula but mode cutoff width calculations remain least accurate. Similar observations may be made of the phase plots of the reflection coefficients given by Yee *et al.* (1968).

By using a Fresnel integral scattering function for the multiply diffracted fields rather than its asymptotic form Felsen and Yee (1968) improved the numerical values obtained by this method, particularly near TEM cutoff where accuracy for guide widths as small as $a = 0.14\lambda$ was attained.

The correct asymptotic solution for the total reflection coefficient, obtained by Boersma (1975b) with the uniform asymptotic theory of diffraction, is

$$\Gamma_{Nn} = \Gamma_{Nn}^{(0)} \left\{ 1 + \frac{\exp(-j\pi/4)}{\sqrt{2\pi ka}} \frac{k(k_N + k_n)}{k_N k_n} S^{\pm}(ka) \right.$$

$$\left. + \frac{1}{2} \left[\frac{\exp(-j\pi/4)}{\sqrt{2\pi ka}} \frac{k(k_N + k_n)}{k_N k_n} S^{\pm}(ka) \right]^2 + 0[(ka)^{-3/2}] \right\} \quad (9.18)$$

for $N + n$ even. The series

$$S^{\pm}(ka) = \pm \sum_{m=1}^{\infty} \frac{(\mp 1)^{m-1} \exp(-jmka)}{m^{3/2}} \quad (9.19)$$

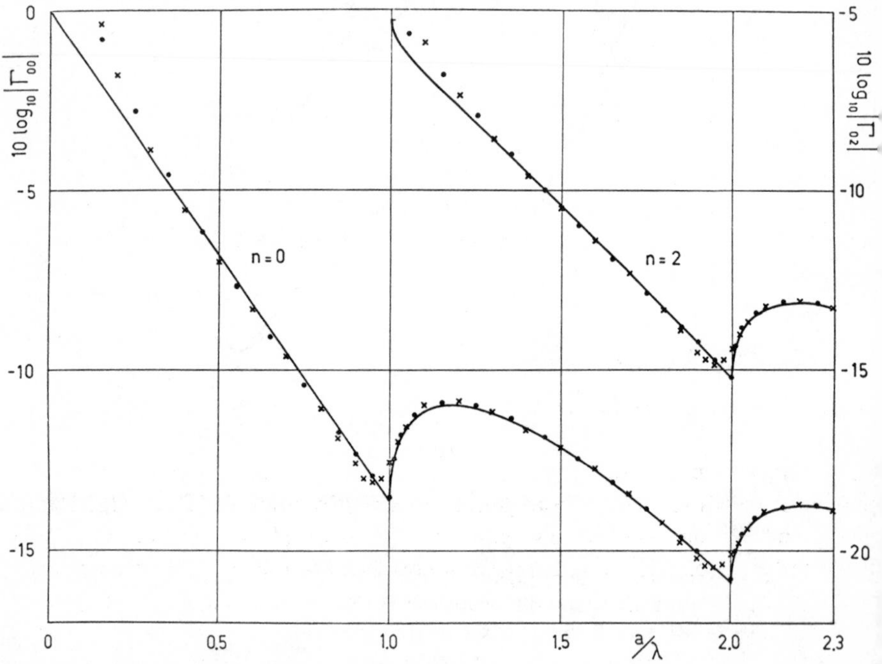

Fig. 9.3 *Reflection coefficients for TEM mode incidence on the open end of a parallel-plate waveguide of width a.*
x x x x   ray-optical theory, eqn. 9.13
. . . . . .   uniform asymptotic theory, eqn. 9.18
————   exact, eqns. 9.15 and 9.16

(Reprinted from Boersma, 1975b)
©1975 SIAM

converges more slowly than eqn. 9.14, but this convergence can be accelerated. Eqn. 9.18 provides numerical values in excellent agreement with the exact results, as illustrated in Fig. 9.3, where it is also compared with eqn. 9.13.

9.2.2 TE polarisation

For a TE-polarised incident field in Fig. 9.1,

$$E_z^i = \sin\left(\frac{N\pi y}{a}\right)\exp(jk_N x), \qquad N = 1, 2, \ldots \tag{9.20}$$

the analysis proceeds similarly with a diffraction coefficient $D(\theta_0, \theta)$ defined by eqn. 8.3 and, for ray to mode conversion with E_z in place of H_z, sin replaces cos in all modal expressions such as eqns. 9.4, 9.6 and 9.8. The reflected field is

$$E_z = \sum_{n=1}^{\infty} R_{Nn} \sin\left(\frac{n\pi y}{a}\right)\exp(-jk_n x) \tag{9.21}$$

with k_n defined by eqn. 9.5.

The noninteraction (single diffraction) reflection coefficient is

$$R_{Nn}^{(0)} = \frac{j\sqrt{k-k_N}\,\sqrt{k-k_n}}{k_n a(k_N + k_n)}, \qquad N + n \text{ even} \tag{9.22}$$

The reflection coefficient R_{Nn} for single and multiple diffraction obtained by Yee *et al.* (1968) with the ray-optical method and by Boersma (1974) with the uniform asymptotic theory are given by eqns. 9.13 and 9.18, respectively, with $\Gamma_{Nn}^{(0)}$ replaced by $R_{Nn}^{(0)}$. Since these TE-reflection coefficients differ from their TM counterparts only by their noninteraction components they behave similarly in relation to Weinstein's exact result, which for $N = n = 1$ is

$$|R_{11}| = \sqrt{\frac{(a/\lambda)-\gamma}{(a/\lambda)-\gamma}}\exp(-\pi\gamma), \qquad 0\cdot 5 < a/\lambda < 1\cdot 5 \tag{9.23}$$

with $\gamma = [(a/\lambda)^2 - 1/4]^{1/2}$.

9.2.3 Flanged waveguides

Numerical results for flanged waveguides, obtained by using the wedge diffraction coefficients of eqn. 8.40 are also given by Yee *et al.* (1968). Lee (1969, 1970) developed a ray-optical method using a modified diffraction coefficient based on a rigorous asymptotic solution for diffraction by two staggered parallel half-planes. Applying it to TEM mode self-reflection from an open-ended parallel plate waveguide with infinite right-angled flanges he obtained a result equal to $(4/\sqrt{27})\Gamma_{00}$, where $|\Gamma_{00}|$ is eqn. 9.15, the exact value for the unflanged guide. The same factor appears to apply to higher order Γ_{0n} reflection coefficients. TE mode reflection coefficients for flanged waveguides are also smaller than their unflanged values but by a factor closer to unity.

9.2.4 Coupling between adjacent guides

Coupling between adjacent parallel-plate waveguides has been calculated by this ray-optical method by Driessen (1979). Single diffraction by the edge of the common guide wall yields the relative amplitude and phase of the nth mode in the coupled guide to the Nth mode in the incident guide

$$C_{Nn}^{(0)} = \frac{-j\epsilon_n}{4k_n a} \frac{\sqrt{k \pm k_N} \sqrt{k \pm k_n}}{k_N + k_n}$$
(9.24)

where the upper signs apply for TM polarisation and the lower signs for TE. This is half the magnitude of eqns. 9.7 and 9.22 as only one edge contributes. For double diffraction there are two ray contributions, for triple three, and so on with the number growing with the order of diffraction. Only singly, doubly and triply diffracted fields are included in the numerical results of Fig. 9.4 for TEM-TEM coupling between adjacent guides of equal widths. There is no exact solution to this problem, but the accuracy of these numerical results is expected to be comparable to those for at least the singly and doubly diffracted field contributions of Fig. 9.2.

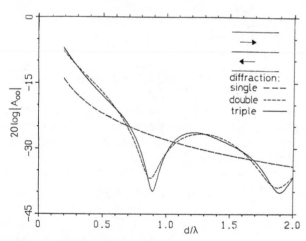

Fig. 9.4   *Amplitude of TEM coupling between adjacent parallel-plate waveguides of equal width d*

— — — — single diffraction only

· · · · · · · single and double diffraction

——————— single, double and triple diffraction

(Reprinted from Driessen, 1979)

9.3 Radiation from open-ended parallel-plate waveguides

9.3.1 TM polarisation

For a TM-polarised incident field given by eqn. 9.1, the singly diffracted field from the lower and upper edges of the aperture may be written as eqn. 9.2 in $x < 0$ of Fig. 9.5. With TEM incidence $N = 0$ and $\theta_N = 0$. In the coordinates of Fig. 9.5 eqn. 9.2 becomes

$$H_z^{(0)} = -\frac{\exp(-j\pi/4)}{2\sqrt{2\pi k}}\left\{\sec\left(\frac{\theta}{2}\right)\frac{\exp(-jkr_1)}{\sqrt{r_1}} + \sec\left(\frac{\theta'}{2}\right)\frac{\exp(-jkr_2)}{\sqrt{r_2}}\right\} \quad (9.25)$$

Behind the aperture plane, $x = 0$, singly diffracted fields from the far edge are blocked by the near half-plane. Consequently in $\pi/2 < \phi < \pi$ only the second term of eqn. 9.25 applies and in $-\pi < \phi < -\pi/2$ the singly diffracted field is represented by the first term of eqn. 9.25. The far-field pattern is obtained by using

$$r_1 = r + \frac{a}{2}\sin\phi, \qquad \theta = \pi - \phi$$
$$\hspace{10.5cm}(9.26)$$
$$r_2 = r - \frac{a}{2}\sin\phi, \qquad \theta' = \pi + \phi$$

giving

$$H_z^{(0)} = \frac{\exp[-j(kr - \pi/4)]}{\sqrt{r\lambda}}\, a\, \frac{\sin[(ka/2)\sin\phi]}{ka\sin(\phi/2)}, \quad |\phi| < \pi/2 \hspace{1.5cm}(9.27)$$

$$= \pm\frac{\exp[-j(kr + \pi/4) \pm j(ka/2)\sin\phi]}{\sqrt{r\lambda}\, 2k\sin(\phi/2)} \quad \left.\begin{array}{l} \pi/2 < \phi < \pi \\ -\pi < \phi < -\pi/2 \end{array}\right\} \hspace{0.6cm}(9.28)$$

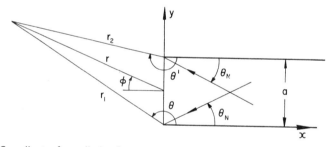

Fig. 9.5 *Coordinates for radiation from an open-ended parallel-plate waveguide*

The diffracted fields from each edge of eqn. 9.25 are individually singular in the axial ($\theta = \theta' = \pi$) direction but added together their singularities cancel and produce the effect of the incident field, leaving eqn. 9.27 well behaved in the $\phi = 0$ direction. The same situation occurs in diffraction by a slit (see Section 8.2.1).

At $\phi = \pm\pi/2$ the singly diffracted field of eqns. 9.27 and 9.28 is discontinuous by a factor $-j\exp(jka/2)[2\sin(ka/2)]^{-1}$. To obtain a continuous pattern across the aperture plane interaction between the edges needs to be included. Yee and Felsen (1968) did this by replacing the edges by equivalent isotropic line sources. For first-order interaction the upper edge is illuminated by an isotropic line source of the form of eqn. 9.10, and the far-field expression of eqn. 7.51 with a (+) sign between the terms for TM polarisation is used. When this doubly diffracted field is added to eqns. 9.27 and 9.28 the pattern discontinuity at $\phi = 0$ is approximately halved, this being the factor by which an incident field is reduced along a shadow boundary (see eqn. 8.6). Higher-order interaction is calculated similarly, each successive interaction further reducing the pattern discontinuity by approximately one-half. In this way the pattern discontinuity can be made arbitrarily small.

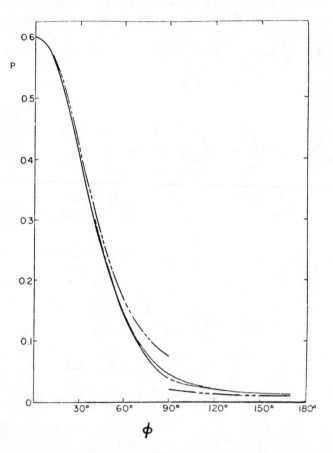

Fig. 9.6   *Radiated power patterns for TEM mode incidence on the open end of a parallel-plate waveguide of width a = 0·6λ*
—·—·—   single diffraction, from eqns. 9.27, 9.28
— ·· —   single and multiple diffraction
————   exact, eqn. 9.29

(Reprinted from Yee and Felsen, 1968)

In Fig. 9.6, the discontinuous power pattern is from the above expressions for the singly diffracted fields while the continuous broken curve includes about seven interactions between the edges. The exact result for the radiated power (Weinstein, 1966, p. 33) is

$$P(\phi) = \frac{Z_0 a^2}{2r\lambda} \frac{\sin[(ka/2)\sin\phi]}{(ka/2)\sin\phi} \exp[-(ka/2)(1-\cos\phi)], \qquad 0 < a < \lambda$$

$$(9.29)$$

and is represented by the solid curve in Fig. 9.6. The main discrepancy is in the vicinity of the aperture plane $\phi = \pm\pi/2$. In the axial direction ($\phi = 0$) interaction fields vanish and the axial power calculated from eqn. 9.27

$$P(0) = \frac{1}{2} Z_0 |H_z^{(0)}(0)|^2 = \frac{Z_0 a^2}{2r\lambda} \qquad (9.30)$$

agrees with eqn. 9.29.

The discrepancy between ray-optical and exact results can be reduced by using the uniform asymptotic theory of diffraction, which accounts for the anisotropic pattern of the edge diffracted fields in interactions between the edges. A simplified version of this calculation was used by Rudduck and Wu (1969) and James (1976) who added to the first interaction fields of the ray optical analysis an additional 'slope diffraction' term similar to eqn. 8.21 to account for the anisotropic edge diffracted field. In this way singly and doubly diffracted fields alone provided a good approximation to the exact pattern. This indicates the major approximation of the ray-optical method is its assumption of diffraction by isotropic line sources. Bowman (1970) showed this from an asymptotic expansion of the exact TE result for scattering by an open-ended waveguide. The singly diffracted field and the first and second interaction fields of the ray-optical analysis agreed with the exact result but subsequent ray-optical interactions underestimated the asymptotic result.

9.3.2 TE polarisation

With TE_1 mode incidence in Fig. 9.5 ($N = 1$ in eqn. 9.20), the singly diffracted fields from the two edges are

$$E_z^{(0)} = \frac{-j}{2} \left\{ D(\theta_1, \theta) \frac{\exp(-jkr_1)}{\sqrt{r_1}} + D(\theta_1, \theta') \frac{\exp(-jkr_2)}{\sqrt{r_2}} \right\} \qquad (9.31)$$

where

$$\theta_1 = \sin^{-1}(\lambda/2a) \qquad (9.32)$$

is the dominant mode angle of incidence on the edges. Using eqn. 8.3 for $D(\theta_1, \theta)$ and the far-field expressions eqn. 9.26, the noninteraction fields are

$$E_z^{(0)} = \frac{\exp[-j(kr - \pi/4)]}{\sqrt{\lambda r}} \left(1 - \frac{k_1}{k}\right)^{1/2} (1 + \cos\phi)^{1/2} \frac{\cos[(ka/2)\sin\phi]}{k\cos\phi - k_1},$$

$$|\phi| < \pi/2 \qquad (9.33)$$

$$= \frac{1}{2} \frac{\exp[-j(kr - \pi/4) + j(ka/2)|\sin\phi|]}{\sqrt{\lambda r}} \left(1 - \frac{k_1}{k}\right)^{1/2} \frac{(1 + \cos\phi)^{1/2}}{k\cos\phi - k_1},$$

$$\pi/2 < |\phi| < \pi \qquad (9.34)$$

where

$$k_1 = k \cos \theta_1 = [k^2 - (\pi/a)^2]^{1/2}$$

In $|\phi| > \pi/2$, only the singly diffracted field from the nearest edge is used, and consequently the radiation pattern is discontinuous by a factor $\exp(jka/2)$ $[2\cos(ka/2)]^{-1}$ at the aperture plane. As with TEM incidence, this discontinuity is

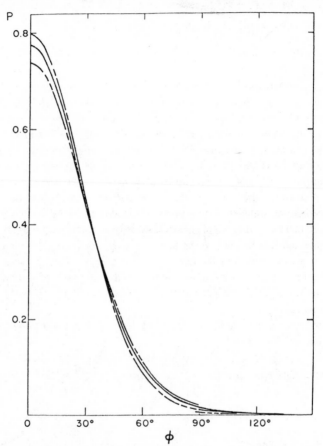

Fig. 9.7 *Radiated power patterns for* TE_1 *mode incidence on the open end of a parallel-plate waveguide of width* $a = 0.8\lambda$

— - - — single diffraction, from eqns. 9.33, 9.34

— · — single and multiple diffraction

——— exact, eqn. 9.35

(The three results agree for $\phi = \theta_1 = \sin^{-1} (\lambda/2a) = 38 \cdot 7°$)

(Reprinted from Yee and Felsen, 1968)

approximately halved by including the doubly diffracted fields occurring from first-order edge interaction, halved again by second-order interaction and so on, and thus the pattern can be made continuous as shown in Fig. 9.7.

The diffracted field terms of eqn. 9.31 are individually singular along the shadow boundaries of the incident waves $\phi = \pm\theta_1$, but together in eqn. 9.33 the singularities cancel and so include the incident field. In these directions the field amplitude from eqn. 9.33 agrees with that obtained from the exact result for the total radiated power (Weinstein, 1966, p. 33)

$$P(\phi) = \frac{Y_0}{4} \frac{k_1 a^2 \exp(-k_1 a/2)}{(1 + k/k_1)r} (1 + \cos\phi) \frac{\cos[(ka/2)\sin\phi]}{\pi^2 - (ka\sin\phi)^2}$$

$$\times \exp[(ka/2)\cos\phi], \qquad \lambda/2 < a < 3\lambda/2 \qquad (9.35)$$

and consequently the radiation field of eqn. 9.33 is most accurate in these directions.

Interesting comparisons between exact and Kirchhoff approximation TM and TE radiation patterns have been made by Weinstein (1966) and Collin and Zucker (1969). James (1976) has added some geometrical diffraction theory patterns. Weinstein (1966) showed that for arbitrary mode incidence the amplitude of the radiated field in the two directions of the incident plane waves in the guide can be determined from Huygen's principle and the Kirchhoff approximation. He states the phase can also be shown to agree in these directions. Yee and Felsen (1968) observed that this exact field amplitude on the shadow boundaries of the incident waves of open-ended parallel plate waveguides is also provided by the geometrical theory of diffraction with single diffraction only at the two aperture edges for arbitrary mode incidence.

9.3.3 Axial gain
For two-dimensional TM fields, aperture gain may be calculated from

$$G = \frac{\frac{1}{2}Z_0 |H_z|^2}{P_r/(2\pi r)} \qquad (9.36)$$

where P_r is the incident power density per unit length in the z-direction. With TEM mode incidence this is

$$P_r = \frac{1}{2} Re \int_0^a E_y^i H_z^{i*} dy = \frac{a}{2} Z_0 \qquad (9.37)$$

and with eqn. 9.30 in eqn. 9.36 the axial gain obtained by the geometrical theory of diffraction is

$$G = ka \qquad (9.38)$$

an exact result also obtainable by the Kirchhoff theory.
For TE_1 mode incidence the incident power density is

$$P_r = \frac{1}{2} \int_0^a E_z^i H_y^{i*} dy = \frac{k_1 a}{4k} Y_0 \qquad (9.39)$$

and the gain is obtained from

$$G = \frac{\frac{1}{2} Y_0 |E_z|^2}{P_r/(2\pi r)} \tag{9.40}$$

The gain in the directions $\phi = \pm\theta_1$ with $E_z^{(0)}$ for $\phi = \theta_1$ from eqn. 9.33 in eqn. 9.40 is $G = k_1 a/2$, an exact result also obtainable by Kirchhoff theory. Geometrical diffraction and Kirchhoff theory axial gains differ however, neither being exact. This provides an opportunity to compare their relative accuracy in the direction where Kirchhoff theory is usually its most reliable. With only single diffraction at the aperture, eqn. 9.33 gives with $\phi = 0$

$$E_z^{(0)} = \frac{a}{\pi}\sqrt{\frac{2}{\lambda r}} (1 + k_1/k) \exp[-j(kr - \pi/4)] \tag{9.41}$$

The effect of double diffraction by singly diffracted rays which cross the aperture is to multiply eqn. 9.41 by the factor

$$1 + \left(\frac{k}{k_1} - 1\right) \sqrt{\frac{2}{\pi}} \exp(j\pi/4) \, F\,[(\sqrt{ka}] \tag{9.42}$$

in which isotropic line sources are assumed. The axial field then includes first-order interaction by the edges. According to previous examples and the aperture field behaviour (Jull, 1973a) similar higher-order interactions are unlikely to provide significant improvements in accuracy.

Using eqn. 9.41 for E_z in eqn. 9.40, the geometrical theory of diffraction result for the axial gain is

$$G^{(0)} = \frac{8a}{\pi\lambda} (1 + \lambda_g/\lambda) \tag{9.43}$$

where

$$\lambda_g = 2\pi/k_1 = \lambda/\sqrt{1 - (\lambda/2a)^2} \tag{9.44}$$

is the wavelength of the incident mode in the guide. Eqn. 9.43 includes only single diffraction at the aperture. With double diffraction, i.e. with eqns. 9.41 and 9.42 in eqn. 9.40, the gain is

$$G^{(1)} = \frac{8a}{\pi\lambda} (1 + \lambda_g/\lambda) \left| 1 + \left(\frac{\lambda_g}{\lambda} - 1\right) \sqrt{\frac{2}{\pi}} \exp(j\pi/4) \, F\,[\sqrt{ka}] \right|^2 \tag{9.45}$$

A Kirchhoff theory solution for the axial field is obtained by using $E_z^i = \sin(\pi y/a)$ as the aperture field in eqn. 2.30 with $\theta = 0$. Let this be

$$E_z^{K1} = \frac{\exp[-j(kr - \pi/4)]}{\sqrt{r\lambda}} \int_0^a \sin\left(\frac{\pi y}{a}\right) dy$$

$$= \frac{2a}{\pi} \frac{\exp[-j(kr - \pi/4)]}{\sqrt{r\lambda}} \tag{9.46}$$

This approximation is based on the tangential electric field in the aperture only. A Kirchhoff solution for the axial field based on both tangential electric and magnetic fields in the aperture is

$$E_z^{K2} = \frac{\exp[-j(kr - \pi/4)]}{2\sqrt{r\lambda}} \int_0^a (E_z^i + Z_0 H_y^i)\, dy$$

$$= \frac{2a}{\pi} \left(1 + \frac{k_1}{k}\right) \frac{\exp[-j(kr - \pi/4)]}{\sqrt{r\lambda}} \tag{9.47}$$

when the aperture magnetic field is $H_y^i = (j\omega\mu_0)^{-1}\partial E_z^i/\partial x = Y_0(k_1/k)E_z^i$. For $a \gg \lambda$, $k_1 \approx k$ and tangential electric and magnetic fields in the aperture are related by free space conditions, as in eqn. 9.46, but relatively small apertures are under consideration here.

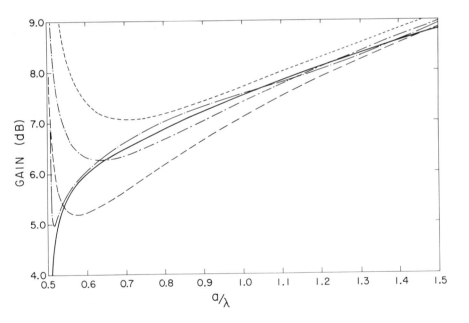

Fig. 9.8 *Axial gain per unit length for* TE_1 *mode incidence on the open end of a parallel-plate waveguide*
Kirchhoff Theory: - - - - eqn. 9.48 — — — — eqn. 9.49; Geometrical theory of diffraction: — - - eqn. 9.43——·—— eqn. 9.45; Exact: ——— eqn. 9.50 ©1973 IEEE (Jull, 1973a).

Using eqns. 9.46 and 9.47 in eqn. 9.40, the Kirchhoff gain values are, respectively,

$$G^{K1} = \frac{16a\lambda_g}{\pi\lambda^2} \tag{9.48}$$

and

$$G^{K2} = \frac{8a}{\pi\lambda}\left(1 + \frac{\lambda_g}{\lambda}\right) \tag{9.49}$$

The exact result, obtained by using eqn. 9.35 with $\phi = 0$ as $\frac{1}{2}Y_0|E_z|^2$ in eqn. 9.40 is

$$G = \frac{8a}{\lambda(1 + \lambda_g/\lambda)}\exp\left[(\pi a/\lambda)(1 - \lambda/\lambda_g)\right], \qquad \lambda/2 < a < 3\lambda/2 \tag{9.50}$$

Numerical values of the gain from these expressions are shown in Fig. 9.8 for the range $0.5 < a/\lambda < 1.5$. Evidently for $a/\lambda > 0.6$ the first Kirchhoff approximation G^{K1} yields values above the correct result whereas G^{K2} yields values below it. That G^{K2} happens to coincide with the exact value for a guide width slightly above cutoff was shown by Weinstein (1966, Fig. 13), but in general the simple geometrical theory of diffraction result $G^{(0)}$ is more accurate than either Kirchhoff approximation. At frequencies near cutoff the inclusion of first-order edge interaction through eqn. 9.45 yields good agreement with the exact result for $a/\lambda > 0.52$. At high frequencies ($\lambda_g \approx \lambda$) the approximations converge.

9.4 Electric line sources

9.4.1 Strip reflector

A direct application of Keller's geometrical theory of diffraction encounters immediate difficulty due to reflection and shadow boundary singularities in the radiation pattern. This can be avoided by using line source, rather than plane-wave, excitation of the edges, e.g. eqn. 7.51, which is a far-field solution free of such singularities.

A simple, practical problem which illustrates this use of geometrical diffraction theory is the calculation of the far-field pattern of an electric dipole parallel to a conducting strip, as in Fig. 9.9. The *H*-plane pattern of an electric line source is identical to that of a dipole. The field of an electric line source over an infinite conducting plane is composed of the far field of the source in isolation

$$E_z^i = \frac{\exp\{-jk[r - r_0\cos(\theta - \theta_0)]\}}{\sqrt{kr}} \tag{9.51}$$

and a reflected far field

$$E_z^r = -\frac{\exp\{-jk[r - r_0\cos(\theta + \theta_0)]\}}{\sqrt{kr}} = -E_z^i \exp(-j2kh\sin\theta) \quad (9.52)$$

which appears to originate from the image of the dipole below the plane. These are the geometrical optics fields in front of the strip and are a crude first approximation to the total fields. A much better approximation also includes diffracted fields from the two edges. The field in front of the strip can be written as the sum

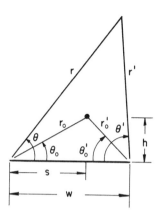

Fig. 9.9 *Electric dipole (or line source) parallel to a strip reflector*

of the total field of a line source over a half plane for each of the edges, less the field of a dipole over an infinite conducting plane. Behind the strip there are no geometrical optics fields and the solution is just the sum of the two half-plane solutions. This total field may be written

$$E_z = \frac{\exp(-jkr)}{\sqrt{kr}} f(\theta, r_0, \theta_0) + \frac{\exp(-jkr')}{\sqrt{kr'}} f(\theta', r_0', \theta_0'), \qquad 0 \leqslant {\theta \atop \theta'} \leqslant 2\pi$$

$$-(E_z^i + E_z^r) \qquad\qquad\qquad\qquad\qquad 0 \leqslant {\theta \atop \theta'} \leqslant \pi$$

$$(9.53)$$

where $f(\theta, r_0, \theta_0)$ is defined by eqn. 7.50.

Alternatively, the diffracted field of an edge E_z^d may be separated from eqn. 7.51 and written as eqn. 7.58 with

$$v(r_0, \alpha) = -\frac{\exp[jkr\cos\alpha + j\pi/4]}{\sqrt{\pi}} F\left[\sqrt{2kr}\left|\cos\frac{\alpha}{2}\right|\right] \qquad (9.54)$$

obtained from eqn. 7.59 with $n = 2$. Then the far field is

$$E_z = E_z^i + E_z^r + E_z^d + E_z^{d'} \qquad 0 \leqslant \theta \leqslant \pi$$

$$= E_z^d + E_z^{d'} \qquad\qquad\qquad \pi \leqslant \theta \leqslant 2\pi$$

$$(9.55)$$

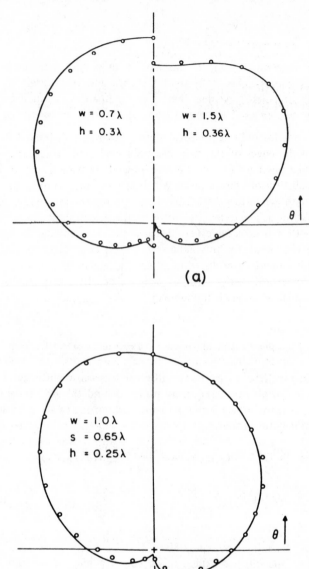

Fig. 9.10   *H-plane patterns of an electric dipole parallel to and at a distance h from a reflecting strip of width w*

   a Symmetrical excitation (*s* = *w*/2)
   b Asymmetrical excitation
      o o o predicted from eqn. 9.53 or 9.55
      —— measured

(Reprinted from VanKoughnett and Wong, 1971)

where

$$E_z^{d'} = \frac{\exp(-jkr')}{\sqrt{kr'}} \, [v(r_0',\theta' - \theta_0') - v(r_0',\theta' + \theta_0')] \tag{9.56}$$

is the diffracted field of the other edge. Eqns. 9.53 and 9.55 give identical results. As they are far-field $(r \gg w)$ expressions, the substitutions $r' \approx r - w \cos\theta$, $\theta' \approx \pi - \theta$ apply.

This result contains only geometrical optics and singly diffracted fields; inter-action between the edges of the strip has been neglected. There are no shadow boundary singularities here for the Fresnel integrals provide a smooth transition. The far field substitutions give a result which is easily programmed for numerical values. Some are shown in Fig. 9.10. There is very good experimental agreement even for strips as narrow as a wavelength. Not even the discontinuity in the calculated patterns at $\theta = 0, \pi$ is noticeable here, because the fields interacting between the strip edges are very weak for this polarisation. Tsai *et al.* (1972) have shown by comparison with numerical results from an integral equation solution of this problem that the geometrical theory of diffraction yields satisfactory results for reflector widths w as small as 0.2λ when double diffraction is included.

9.4.2 Corner reflector

Another early example of the application of geometrical diffraction theory to antennas is the analysis of the *H*-plane pattern of a corner reflector by Ohba (1963). Tsai *et al.* (1972) included as well a very useful comparison with results from the numerical solution of an integral equation. In Fig. 9.11, the source is parallel to the intersection of two conducting planes and at a distance s along the bisector of the $90°$ corner. A conventional analysis is by the method of images by which the pattern is that of a ring array of four sources of equal magnitude and alternating polarity (e.g. Kraus, 1950, p. 329). For a line source feed the field is

$$E_z^g = \sum_{n=1}^{4} (-1)^{n+1} \frac{\exp(-jkr_n)}{\sqrt{kr_n}}, \qquad |\phi| < \pi/4 \tag{9.57}$$

Insertion of the far field distances r_n of the source and images to the field point

$$\left. \begin{array}{l} r_1 = r - s\cos\phi \\ r_2 = r - s\sin\phi \\ r_3 = r + s\sin\phi \\ r_4 = r + s\cos\phi \end{array} \right\} \tag{9.58}$$

in the phase terms of eqn. 9.57 and replacing r_n by r in the amplitude terms gives the radiation pattern of the geometrical optics field.

$$E_z^g = 2\frac{\exp(-jkr)}{\sqrt{kr}} \, [\cos(ks\cos\phi) - \cos(ks\sin\phi)] \qquad |\phi| < \pi/4 \tag{9.59}$$

This results in axial maxima for dipole to corner spacings of $s = \lambda/2, 3\lambda/2, \ldots$ for infinite reflectors.

With finite reflector walls as in Fig. 9.11, eqn. 9.59 applies only for $|\phi| < \pi/4 - \theta_0$. Since the pattern is symmetrical about $\phi = 0$ it suffices to consider only the angular range $0 \leqslant \phi \leqslant \pi$. In $\phi > \pi/4 - \theta_0$, $\pi/4 - \theta_0'$ and $\pi/4 + \theta_0'$ images 2, 3 and 4, respectively, do not contribute and in $\phi > \pi/4 + \theta_0$ there is no geometrical optics field.

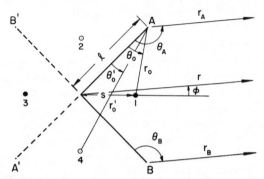

Fig. 9.11 *90° corner reflector antenna*
The electric line source or dipole is at 1 and its images at 2, 3 and 4. A′ and B′ are images of the reflector edges A and B

Everywhere there are diffracted fields from at least one of the reflector edges A and B which may be included through the expressions for line source diffraction of the source and its images by the two half planes which form the reflector. Images A′ and B′ of the edges are introduced to include reflection after diffraction. For example, let

$$v(r_A, r_0, \alpha) = \frac{\exp(-jkr_A)}{\sqrt{kr_A}} v(r_0, \alpha) \tag{9.60}$$

where $v(r_0, \alpha)$ is defined by eqn. 9.54. Then the singly diffracted fields of the outer edges in $0 < \phi < \pi/4 - \theta_0$ may be written in the coordinates of Fig. 9.11 as

$$\begin{aligned}
E_z^{sing.diff.} &= v(r_A, r_0, \theta_A - \theta_0) - v(r_A, r_0, \theta_A + \theta_0) \\
&+ v(r_B, r_0, \theta_B - \theta_0) - (r_B, r_0, \theta_B + \theta_0) \\
&- v(r_A, r_0', \theta_A - \theta_0') + v(r_A, r_0', \theta_A + \theta_0') \\
&- v(r_B, r_0', \theta_B - \theta_0') + v(r_B, r_0', \theta_B + \theta_0') \\
&- v(r_{A'}, r_0, \theta_A - \theta_0) + v(r_{A'}, r_0, \theta_A + \theta_0) \\
&+ v(r_{A'}, r_0', \theta_A - \theta_0) - v(r_{A'}, r_0', \theta_A + \theta_0').
\end{aligned} \tag{9.61}$$

The first eight terms in eqn. 9.61 account for diffraction by the source and its images by the edges without subsequent reflection. The remaining terms, which contain the distance $r_{A'}$ from the image of edge A to the field point, include reflection after diffraction.

For $\phi = 0$, eqn. 9.61 simplifies as $r_A = r_B$ and $\theta_A = \theta_B$. Also diffraction plus reflection contributions equal to the last four terms of eqn. 9.61 are added. In $\phi >$ $\pi/4$ there is no reflection plus diffraction and the last four terms of eqn. 9.61 are excluded. In the region $\pi/2 < \phi < \pi$ contributions from edge B are blocked, leaving the fields of only the first, second, fifth and sixth terms of eqn. 9.61. In the rear direction ($\phi = \pi$) the first eight terms contribute. Shadow boundaries of the source and its images are at $\phi = \pi/4 \pm \theta_0$, and $\pi/4 \pm \theta_0'$, but all terms of eqn. 9.61 are well behaved in these directions.

Ohba (1963) showed how doubly diffracted fields could be included. For his example with $l = \lambda$, $s = 0 \cdot 3\lambda$, these were small enough to be negligible. By comparison with a numerical solution of integral equations Tsai *et al.* (1972) showed this method of analysis provides good results for two-dimensional 90° corner reflectors as small as $l = 0 \cdot 25\lambda$, $s = 0 \cdot 3\lambda$. Accurate prediction of the *H*-plane pattern backlobes of a corner reflector of finite length in the *z*-direction requires inclusion of fields diffracted by all reflector aperture edges. Omission of contributions from edges orthogonal to the dipole feed limits the accuracy of these expressions in the rear directions, as is evident in Ohba's (1963) results.

As reflectors with corner angles equal to π/N, where *N* is an integer, have $2N + 1$ images, corner angles smaller than $\pi/2$ introduce more diffraction terms. Arbitrary corner angles require a more general and less convenient image scheme.

9.4.3 Parabolic cylinder reflector

For reflectors larger than a few wavelengths a parabolic reflector as in Fig. 9.12 provides higher directivity than a corner reflector. Ohba (1962) and Lewin (1972) used the geometrical theory of diffraction to obtain the radiation pattern in the lateral and rear directions. As the axis is a confluence of reflected rays, ray methods cannot be conveniently used to obtain the pattern on or near the beam axis. The Kirchhoff method will provide an accurate solution for axial and near axial fields which merges with the geometrical diffraction theory solution well off the beam axis.

To calculate the radiation pattern of a dipole along the focus of a parabolic cylinder, as in Fig. 9.12, the reflector edges are replaced by inclined half-planes tangent to the parabolic surface at the edge. These half-planes are inclined at θ_0 to the reflector axis where

$$\theta_0 = \frac{\pi - \psi_0}{2} = \tan^{-1}(4f/d) \tag{9.62}$$

in which $2\psi_0$ is the angle subtended by the reflector at the feed, f is the focal length and d the aperture width.

As the pattern is symmetrical about $\phi = 0$ only the region $0 < \phi < \pi$ is considered. Except on the axis the only geometrical optics field is the far field of the line source at the focus E_z^i given by eqn. 9.51. Singly diffracted fields from the two edges which are not reflected from the interior surface of the reflector are

$$E_z^d = \frac{\exp(-jkr)}{\sqrt{kr}} [v(r_0, \theta - \theta_0) - v(r_0, \theta + \theta_0)] \tag{9.63}$$

from the upper edge and from the lower edge

$$E_z^{d'} = \frac{\exp(-jkr')}{\sqrt{kr'}}[v(r_0,\theta'-\theta_0)-v(r_0,\theta'+\theta_0)] \tag{9.64}$$

where $v(r_0,\alpha)$ is defined by eqn. 9.54.

Reflection of diffracted fields from the concave surfaces of the parabolic cylinder are less easily accounted for as multiple reflection occurs. In an analysis based on edge diffracted fields reflected from a circularly cylindrical reflector Kinber (1961) approximated these multiply reflected fields by a modified singly reflected ray field. Since then more complete analysis of ray reflection from concave cylindrical surfaces have been given by Babich and Buldyrev (1972) and Ishihara *et al.* (1978). Here these singly diffracted fields contribute only in the angular sector $\phi < \theta_0$. With their omission a geometrical diffraction theory solution for the pattern, applicable mainly in the lateral and rear directions, is

$$\begin{aligned}
E_z &= E_z^i + E_z^d + E_z^{d'}, & \theta_0 &< \phi < \pi/2 \\
&= E_z^i + E_z^d, & \pi/2 &< \phi < 2\theta_0 \\
&= E_z^d, & 2\theta_0 &< \phi < \pi-\theta_0 \\
&= E_z^d + E_z^{d'}, & \pi-\theta_0 &< \phi < \pi
\end{aligned} \tag{9.65}$$

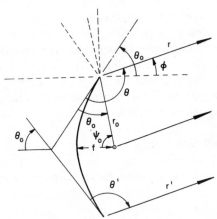

Fig. 9.12 *Parabolic cylinder reflector antenna with a line source or dipole feed at the focus*
Dashed lines indicate transition regions of the diffracted fields.

As reflector apertures are usually several wavelengths wide, doubly diffracted fields are much weaker than the above fields and useful mainly to reduce pattern discontinuities at the boundaries between the above angular sectors and to provide details in the pattern behind the reflector.

To calculate the pattern in the forward sector $0 \leqslant \phi < \theta_0$, Ohba (1963) approximated the parabolic surface with flat ribbon plates. Their geometrical diffraction theory contributions summed agreed closely with the Kirchhoff theory result for the axial and near-axial field. Also, in comparing with experimental values the direct wave from the dipole was suppressed, as is usual in antenna design.

For a circular paraboloidal reflector with a small prime focus feed only two edge diffracted rays are required to obtain the off-axis field but a vector diffraction coefficient is required. James and Kerdemelides (1973) used equivalent edge currents to obtain the pattern and Rusch and Sørensen (1975) the geometrical theory of diffraction. Kouyoumjian (1975) used dyadic diffraction coefficients, double diffraction and inclusion of singly diffracted fields reflected once from the concave surface of the reflector. There the angular sector in which geometrical diffraction theory fails in the forward direction is given as $\phi < 10(kd)^{-1}$ radians.

9.5 Magnetic line sources

9.5.1 Slot in a conducting half-plane

A narrow slot in a conducting plane radiating with the electric field applied across the slot can be represented by a magnetic line source on the conductor. The pattern of a slot parallel to the edge of a wedge or half-plane can be obtained from eqn. 7.60 or the exact solution for the total far field of a magnetic line source near a conducting half plane, the TM equivalent of eqn. 7.51. In the coordinates of Fig. 7.6 this is

$$H_z(r,\theta) = \frac{\exp[-j(kr - \pi/4)]}{\sqrt{\pi kr}} \left\{ \exp[jkr_0 \cos(\theta - \theta_0)] \, F\left[-\sqrt{2kr_0}\right. \right.$$
$$\left. \times \cos\left(\frac{\theta - \theta_0}{2}\right)\right] + \exp[jkr_0\cos(\theta + \theta_0)] \, F\left[-\sqrt{2kr_0}\right.$$
$$\left.\left. \times \cos\left(\frac{\theta + \theta_0}{2}\right)\right] \right\} \tag{9.66}$$

For a magnetic line source on the conductor ($\theta_0 = 0$) at $r_0 = l$, as in Fig. 9.13a, source and image terms in eqn. 9.66 coalesce and a factor of one-half is required. Then the E-plane pattern of the slot is given by

$$H_z(r,\theta) = \frac{\exp[-jk(r - l \cos\theta)]}{\sqrt{kr}} - \frac{\exp[-jk(r - l \cos\theta) + j(\pi/4)]}{\sqrt{\pi kr}}$$
$$\times F[\sqrt{2kl}\cos\theta/2] \tag{9.67}$$

Alternatively,

$$H_z(r,\theta) = H_z^i(r,\theta) + H_z^d(r,\theta) \qquad 0 \leqslant \theta \leqslant \pi$$
$$= H_z^d(r,\theta) \qquad\qquad \pi \leqslant \theta \leqslant 2\pi \tag{9.68}$$

where $H_z^i(r,\theta)$ the first term in eqn. 9.67, is the field of the line source alone and

$$H_z^d(r,\theta) = \frac{\exp(-jkr)}{\sqrt{kr}} v(l,\theta) \qquad (9.69)$$

with $v(l,\theta)$ defined by eqn. 9.54, is the diffracted field of the edge.

Some numerical results (Wait, 1953) are shown in Fig. 9.14. These are obtained from cylindrical mode expansions of the field as in Section 7.7.1, but with the Fresnel integrals, eqn. 9.67 should require less numerical effort.

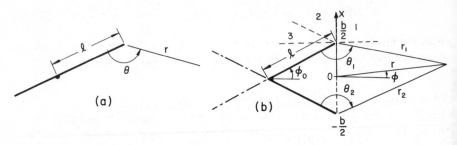

Fig. 9.13 *Magnetic line sources*
  a On a conducting half-plane
  b At the intersection of two half-planes

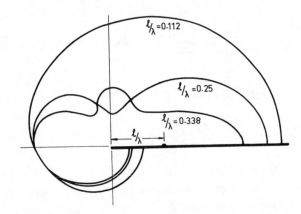

Fig. 9.14 *E-plane patterns of a magnetic line source or radiating slot on a conducting half-plane parallel to the edge*

(After Wait, 1953)

9.5.2 Radiation from pyramidal and sectoral horns

A two-dimensional E-plane sectoral horn is formed by two intersecting half-planes inclined at $2\phi_0$ to each other with a line source at the apex, as in Fig. 9.13b. It serves as a model for the E-plane pattern of pyramidal and sectoral horns. The inclusion of a small parallel-plate feed has only a minor effect on the radiation pattern. Since the field is symmetrical in ϕ only the range $0 \leqslant \phi \leqslant \pi$ need be considered. Neglecting interaction between the horn walls, the field in front of the horn (region 1) can be written as the solution to the line source on a half-plane plus the diffracted field from the lower edge. In region 2 behind the aperture plane the diffracted fields of the lower edge are shadowed by the upper horn wall. In the region 3 directly behind the horn both edges contribute to the pattern. In the notation of eqns. 9.67–9.69, the noninteraction field is

$$
\begin{aligned}
H_z &= H_z(r_1,\theta_1) + H_z^d(r_2,\theta_2) && \text{in 1,} && 0 \leqslant \phi \leqslant \pi/2 \\
&= H_z(r_1,\theta_1), && \text{in 2,} && \pi/2 \leqslant \phi \leqslant \pi - \phi_0 \\
&= H_z^d(r_1,\theta_1) + H_z^d(r_2,\theta_2), && \text{in 3,} && \pi - \phi_0 \leqslant \phi \leqslant \pi \quad (9.70)
\end{aligned}
$$

With the far-field $(r \gg b)$ substitutions $\theta_1 = \pi - \phi_0 + \phi$, $\theta_2 = \pi - \phi_0 - \phi$, and the phase factors $r_1 = r - b/2 \sin \phi$, $r_2 = r + b/2 \sin \phi$, the field in region 1 is

$$
\begin{aligned}
H_z &= \frac{\exp(-jkr)}{\sqrt{kr}} \Bigg\{ \exp[j(kb/2)\sin\phi - jkl\cos(\phi-\phi_0)] \\
&\quad \times \left(1 - \frac{\exp[j(\pi/4)]}{\sqrt{\pi}} F\left[2kl \sin\left(\frac{\phi_0 - \phi}{2}\right) \right] \right) \\
&\quad - \frac{\exp[-j(kb/2)\sin\phi - jkl\cos(\phi+\phi_0) + j(\pi/4)]}{\sqrt{\pi}} \\
&\quad \times F\left[\sqrt{2kl} \sin\left(\frac{\phi_0 + \phi}{2}\right) \right] \Bigg\}
\end{aligned} \quad (9.71)
$$

and in region 2 the second term in the curly brackets is dropped. In region 3

$$
\begin{aligned}
H_z &= -\frac{\exp[-jkr + j(\pi/4)]}{\sqrt{\pi kr}} \Bigg\{ \exp[j(kb/2)\sin\phi - jkl\cos(\phi-\phi_0)] \\
&\quad \times F\left[\sqrt{2kl} \sin\left(\frac{\phi - \phi_0}{2}\right) \right] + \exp[-j(kb/2)\sin\phi - jkl\cos(\phi+\phi_0)] \\
&\quad \times F\left[\sqrt{2kl} \sin\left(\frac{\phi + \phi_0}{2}\right) \right] \Bigg\}
\end{aligned} \quad (9.72)
$$

This is a two-dimensional solution, but these expressions also apply to the E-plane patterns of pyramidal or sectoral horns (Russo *et al.*, 1965). They are compared with measurements in Figs. 9.15a and 9.16. There is good agreement there over most of the range $\phi < \pi/2$, but discontinuities in the pattern at the transition boundaries $\phi = \pi/2$, $\pi - \phi_0$. Also, while details of the pattern behind the horn are not represented, the general level of back radiation is predicted. Similar observations were made from the pattern of the singly diffracted fields alone plotted for the parameters of Fig. 9.15c and 9.15d. The numerical data of Yu *et al.* (1966) in

Fig. 9.16 supports these observations. Pattern asymmetry in the dashed curves of Figs. 9.15 is due to the measuring arrangement.

Interaction between the aperture edges and the horn is included here by the procedure described by Yu *et al.* (1966). Kinber and Popichenko (1972) used a self-consistent method. The strongest interaction fields are from singly diffracted rays

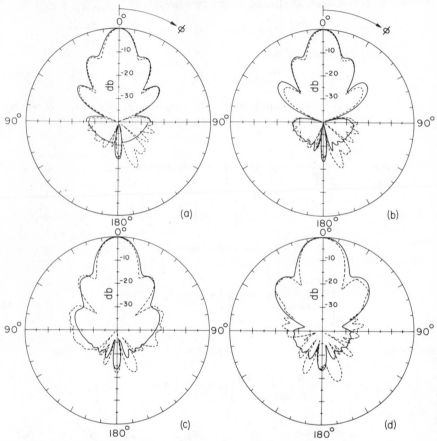

Fig. 9.15 *E-plane radiation patterns of a horn with 1 = 42·15 cm, 2φ₀ = 33°*

$2\phi_0 = 33°$

———— calculated

- - - - measured for a sectoral horn with aperture dimensions 24·00 ×7·21 cm²

a and *b* λ = 7·50 cm

c λ = 9·67 cm

d λ = 11·31 cm

Double diffraction and reflection from the horn interior is included in the predicted patterns of b), c) and d) but not a).

reflected from the interior horn walls into the forward direction, as indicated in Fig. 9.17. To account for them images of the horn edges in the upper and lower walls of the horn are introduced. The number of images in a horn wall is the largest integer less than $\pi/(2\phi_0)$, where $2\phi_0$ radians is the total flare angle of the horn. The far-field contribution from the nth image in the lower wall is

$$H_z^d = \frac{\exp\{-jk\,[(r-(b/2)\sin\phi + l_n\sin(\phi + n\phi_0)]\,\}}{\sqrt{kr}}$$

$$v[l, \pi - (2n+1)\phi_0 - \phi] \tag{9.73}$$

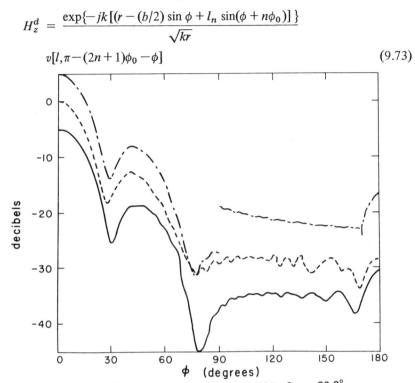

Fig. 9.16 *E-plane patterns of a pyramidal horn with $1_E = 5\cdot61\lambda$, $2\phi_0 = 20\cdot2^\circ$*
Patterns are shifted 5dB for comparison
— · —— no interaction between edges and walls of horn; - - - - with first order interaction; —— measured pattern. (After Yu *et al.*, 1966) © 1966 IEEE

which contributes in the angular region

$$\pi/2 - (n+1)\phi_0 < \phi < \pi/2 - n\phi_0, \qquad n = 1,2,\dots \tag{9.74}$$

In eqn. 9.73, $v(r, \alpha)$ is given by eqn. 9.54 and the distance between the upper edge and its nth image is

$$l_n = l_{n-1}\cos\phi_0 + l_0\cos(n\phi_0) \tag{9.75}$$

with $l_0 = b$, the aperture width. Images in the upper wall contribute mainly in $\phi < 0$. The contribution from the last image may be interrupted by the waveguide feed. If not, and $\pi/(2\phi_0)$ is not an integer, the angular sector in which the last image contributes is

$$0 \leqslant \phi < \pi - (2n+1)\phi_0 \tag{9.76}$$

Inclusion of the above fields does not substantially alter the forward patterns in the numerical samples of Yu *et al.* (1966) for pyramidal horns. In the examples of Fig. 9.15, agreement between predicted and experimental values is less with these diffracted and reflected fields because the measured patterns are for an E-plane sectoral horn and reflection in the horn is poorly approximated by a two-dimensional model. Doubly diffracted fields from the aperture edges can improve the pattern in the

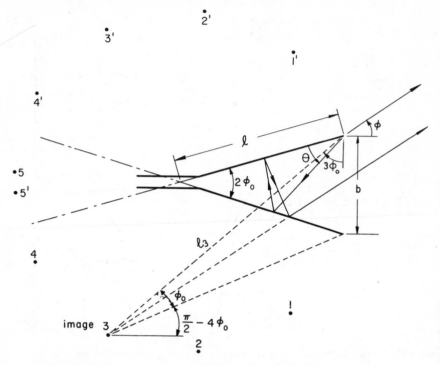

Fig. 9.17 *Images of the edges of a two-dimensional sectoral horn with $2\phi_0 = 33°$*
Diffracted rays from the upper edge which are reflected 3 times and which are un-reflected are shown. The edge images 1 - 4 in the lower wall contribute to the pattern in $\phi > 0$ and the upper wall images 1' - 4' in $\phi < 0$. Ray paths from images 5 and 5' are interrupted by the waveguide feed

lateral and rear directions. Doubly diffracted aperture fields are produced by singly diffracted rays from the aperture edges in the directions

$$\theta_n = \pi/2 - (n+1)\phi_0, \qquad n = 0,1,2,\ldots m \qquad (9.77)$$

For $n = 0$, these rays cross the aperture and are diffracted at the opposite edge where $r = b = 2l \sin \phi_0$, as illustrated in Fig. 9.18. Rays for $n > 0$ are reflected from the horn interior n times before being diffracted again at the same edge (n odd) or the opposite aperture edge (n even). The ray path length between single and double diffraction is

$$d_n = 2l \sin[(n+1)\phi_0] \qquad n = 0,1,2,\ldots m \qquad (9.78)$$

and the angle of incidence for double diffraction is θ_n. m, the maximum value of n, is the largest integer is less than $\pi/(2\phi_0) - 1$ so that θ_m from eqn. 9.77 is always positive. The doubly diffracted far fields of the upper edge may be written as

$$H_z^{dd} = \frac{\exp[-jk(r - (b/2)\sin\phi)]}{\sqrt{kr}} \sum_{n=0}^{m} v\,[l,\pi/2 - (n+1)\phi_0]\,[v(d_n,\pi/2$$

$$+ n\phi_0 + \phi) + v(d_n, 3\pi/2 - (n+2)\phi_0 + \phi)] \qquad (9.79)$$

where $v(r,\alpha)$ is defined by eqn. 9.54.

The upper edge diffracted fields of eqn. 9.79 contribute in $0 \leqslant \phi \leqslant \pi$. The doubly diffracted fields from the lower edge, which may be written as

$$H_z^{dd\prime} = \frac{\exp[-jk(r + (b/2)\sin\phi)]}{\sqrt{kr}} \sum_{n=0}^{m} v\,[l,\pi/2 - (n+1)\phi_0]$$

$$\times\,[v(d_n,\pi/2 + n\phi_0 - \phi) + v(d_n, 3\pi/2 - (n+2)\phi_0 - \phi)] \qquad (9.80)$$

contribute in $0 \leqslant \phi \leqslant \pi/2$ and in $\pi - \phi_0 \leqslant \phi \leqslant \pi$.

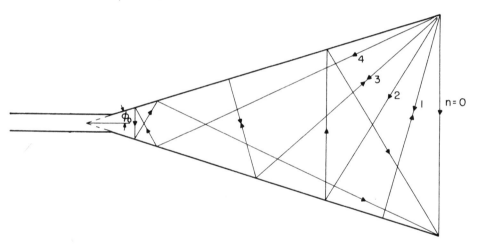

Fig. 9.18 *Ray paths of upper edge diffracted fields between single and double diffraction in a sectoral horn with $2\phi_0 = 33°$*

The calculated patterns (solid curves) of Figs. 9.15c–d and Fig. 9.16 were obtained from a superposition of these doubly diffracted fields and the singly diffracted and geometrical optics fields of eqns. 9.72 and 9.73. In Fig. 9.15, they yield a beamwidth slightly broader than the measured beamwidth of the E-plane sectoral horn. The pattern in the lateral and rear directions is well predicted in Fig. 9.15b but this accuracy decreases as the frequency in increased, as seen in Figs. 9.15c and d. This deterioration in accuracy is less evident for pyramidal horns where the two-dimensional model holds better. Fig. 9.16 shows results for the smallest horn example reported by Yu *et al.* (1966). Safaai-Jazi and Jull (1977) applied the above expressions to a pyramidal horn approximately half this size and obtained good agreement with experiment. The waveguide feed is not accounted for here and its size relative to the aperture affects the accuracy of the predicted patterns.

To calculate the H-plane pattern of a pyramidal horn Yu and Rudduck (1969) decomposed the incident TE_1 mode field of a two-dimensional H-plane sectoral horn into two plane waves which are diffracted at the waveguide-horn junction, reflected from the interior walls of singly diffracted rays is thus much larger than required in the foregoing model. An earlier ray diffraction analysis of radiation from horns by Kinber (1962), which excluded diffraction at the waveguide horn junction, has a similar ray complexity. Mentzer *et al.* (1975) simplified the H-plane pattern prediction of a horn by replacing the waveguide feed with an electric line source at the horn apex and using a slope diffraction coefficient. Singly diffracted fields then provided an H-plane pattern in good agreement with experiment in the forward sector for a pyramidal horn of dimensions typical of gain standards.

In a pyramidal or sectoral horn the H-plane edges, those parallel to the incident electric field, are relatively weakly excited compared to the other edge pair. As a consequence of oblique incidence diffracted fields from the E-plane edges contribute significantly to the H-plane pattern of a horn, particularly in the rear directions where they dominate. Yu and Rudduck (1969) and Mentzer *et al.* (1975) used the method of equivalent edge currents to supplement their geometrical diffraction theory results and complete the H-plane pattern prediction for pyramidal horns. Mentzer *et al.* (1975) also showed that the Kirchhoff method, which is accurate for most forward directions, plus the fields of the equivalent currents on the E-plane edges, also provides a satisfactory approximation to the complete H-plane pattern of a pyramidal horn.

9.5.3 Comparison with the Kirchhoff result

To illustrate the relationship between geometrical diffraction and Kirchhoff theory methods for horn antennas, the on-axis far field of a two-dimensional E-plane sectoral horn is derived here by the Kirchhoff method, for the Kirchhoff approximation is usually at its best for axial fields. At the apex of the horn in Fig. 9.13b there is a magnetic line source of the form

$$H_z^i = \sqrt{\frac{\pi}{2}} \exp[-j(\pi/4)] \, H_0^{(2)}(k\rho) \simeq \frac{\exp(-jk\rho)}{\sqrt{k\rho}}, \qquad k\rho \gg 1 \qquad (9.81)$$

where ρ is the field point distance from the apex. In the aperture plane the approximation of eqn. 6.11 is used, i.e.

$$\rho = \sqrt{(l \cos \phi_0)^2 + x^2} \simeq l \cos \phi_0 + x^2/(2l \cos \phi_0)$$

for $x \ll l \cos \phi_0$. The on-axis far field is, from eqn. 2.29 with $H_z(x,0)$ replacing $E_x(x,0)$ and assumed zero for $|x| > b/2$,

$$H_z(r,0) = \frac{\exp[-j(kr + \pi/4)]}{\sqrt{r\lambda}} \int_{-b/2}^{b/2} H_z(x,0) dx$$

With eqn. 9.81 and the above approximation for ρ, this becomes

$$H_z(r,0) = \frac{\exp[-jk(r + l\cos\phi_0) + j(\pi/4)]}{\sqrt{r\lambda k l\,\cos\phi_0}} \int_{-b/2}^{b/2} \exp[-jkx^2/(2l\cos\phi_0)]dx$$

$$= \sqrt{\frac{2}{kr}}\,\exp[-jk(r + l\cos\phi_0) + j(\pi/4)]\,[C(w) - jS(w)] \quad (9.82)$$

where the Fresnel integrals are those of eqn. 4.8 and

$$w = b/\sqrt{2\lambda l\cos\phi_0} \tag{9.83}$$

$\phi = 0$ in eqn. 9.71 gives for the geometrical diffraction theory result

$$H_z = \frac{\exp[-jk(r + l\cos\phi_0)]}{\sqrt{kr}}\left\{1 - 2\frac{\exp[j(\pi/4)]}{\sqrt{\pi}}\,F\left[\sqrt{2kl}\sin\left(\frac{\phi_0}{2}\right)\right]\right\}$$

$$= \frac{2}{\sqrt{\pi kr}}\,\exp[-jk(r + l\cos\phi_0)]\int_0^{\sqrt{2kl}\sin(\phi_0/2)}\exp(-j\tau^2)\,d\tau$$

$$= \sqrt{\frac{2}{kr}}\,\exp[-jk(r + l\cos\phi_0) + j(\pi/4)]\,[C(w) - jS(w)] \tag{9.84}$$

with

$$w = \frac{b}{\sqrt{2\lambda l}\,\cos(\phi_0/2)} \tag{9.85}$$

Eqns. 9.82 and 9.84 are essentially the same if allowance is made for the small angle approximation in the Kirchhoff approach to this problem. Where it is valid, the Kirchhoff solution includes the geometrical optics fields and the equivalent of the singly diffracted fields which are not reflected from the horn interior. There is little difference, for example, between the predicted radiation pattern of Fig. 9.15a and that obtained from eqn. 6.19 for most of the range of $|\phi| < 90°$. The Kirchhoff result, represented by the broken curve of Fig. 9.19, differs significantly from the geometrical diffraction theory result of Fig. 9.15a only for $\phi > 65°$. Both provide a better approximation to the measured E-plane sectoral horn pattern (dashed curves) for smaller angles off the beam axis than is obtained from the geometrical diffraction theory solution which includes reflection from the horn interior and double diffraction (solid curves of Figs. 9.15b and 9.19). This observation also applies to the results of Figs. 9.15c and d but may not hold for pyramidal horns generally. That the Kirchhoff result contains the geometrical optics result and the equivalent of the singly diffracted fields from the aperture edges is, however, a general result, as shown by Keller *et al.* (1957).

9.5.4 Gain of a two-dimensional E-plane sectoral horn

As pyramidal horns are used as microwave gain standards the effect of diffracted fields from the aperture edges which are reflected from the interior walls or doubly diffracted in the axial direction can be significant. This interaction between horn edges and walls is strongest in the E-plane of the horn and can be approximated by a two-dimensional analysis.

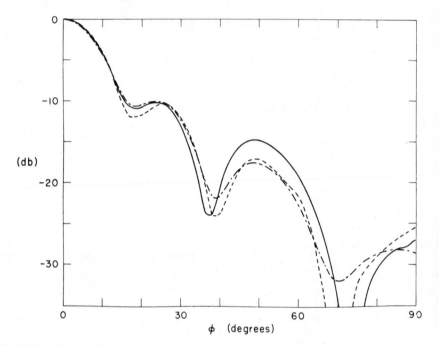

Fig. 9.19   *E-plane radiation patterns of a sectoral horn with $l = 5 \cdot 63\lambda$, $2\phi_0 = 33°$*
——— Geometrical diffraction theory, including double diffraction
—·— Kirchhoff theory, eqn. 6.19
- - - - measured, for an E-plane sectoral horn with aperture dimensions $3 \cdot 20\lambda \times 0 \cdot 96\lambda$

In Fig. 9.17, the last or mth image source contributes to the axial field in accordance with eqns. 9.73 and 9.74 if $\theta_m = \pi - (2m + 1)\phi_0 > 0$ and its ray path is not interrupted by the waveguide feed. This is the direction for edge diffracted fields which proceed in the forward direction after n reflections from the horn interior. The distance along this ray path from the edge to the aperture plane is

$$s_n = l_n \sin(n\phi_0) = 2l \sin \phi_0 \sum_{i=1}^{m} \sin(i\phi_0) \qquad (9.86)$$

Adding $2H_z^d [r + s_n, \pi - (2m + 1)\phi_0]$, where H_z^d is given by eqns. 9.69 to 9.71, the axial geometrical optics and singly diffracted far fields are

$$= \frac{\exp(-jkr)}{\sqrt{kr}} \{ \exp(-jkl \cos \phi_0) + 2v(l, \pi - \phi_0)$$

$$+ 2\exp(-jks_n) \, v(l, \pi - (2m + 1)\phi_0) \} \tag{9.87}$$

Doubly diffracted fields from the two edges in the axial direction are given by twice eqn. 9.79 with $\phi = 0$, i.e. by $2(kr)^{-1/2} \exp(-jkr)S$, where

$$S = \sum_{n=0}^{m} v[l, \pi/2 - (n + 1)\phi_0] \{ v(d_n, \pi/2 + n\phi_0) + v[d_n, 3\pi/2 - $$

$$(n + 2)\phi_0] \} \tag{9.88}$$

If doubly diffracted fields can be reflected from the horn interior into the forward direction without interruption by the waveguide feed they may be included by adding

$$S' = \exp(-jks_n) \sum_{n=0}^{m} v[l, \pi/2 - (n + 1)\phi_0] \{ v[d_n, \pi/2 - (2m - n)\phi_0] + $$

$$+ v[d_n, 3\pi/2 - (2m + n + 2)\phi_0] \} \tag{9.89}$$

to eqn. 9.88.

Higher-order diffracted fields may be similarly included but will be of little consequence for typical horns.

The axial gain per unit length in the z-direction can be obtained from eqn. 9.36 with P_r, the incident power calculated from the incident magnetic feed H_z^i, given by eqn. 9.81, and the incident electric field

$$E_\phi^i = -\frac{1}{j\omega\epsilon_0} \frac{\partial H_z^i}{\partial \rho} = \frac{k}{\omega\epsilon_0} \sqrt{\frac{\pi}{2}} \exp(-j3\pi/4) H_1^{(2)}(k\rho) \tag{9.90}$$

Then

$$P_r = \frac{1}{2} \operatorname{Re} \int^{\phi} E_\phi^i H_z^i \rho \, d\phi$$

$$= \frac{\pi k\rho\phi_0}{2\omega\epsilon_0} \operatorname{Re}[-jH_1^{(2)}(k\rho)H_0^{(2)}(k\rho)] = \frac{\phi}{\omega\epsilon_0} \tag{9.91}$$

If eqns. 9.84 and 9.91 are used in eqn. 9.36, the axial gain for single diffraction (no interaction) at the aperture is

$$G = \frac{2\pi}{\phi_0} [C^2(w) + S^2(w)] \tag{9.92}$$

where w is given by eqn. 9.85. For small flare angles $\phi_0 \approx b/(2l)$ and eqn. 9.92 becomes

$$G = kbR_E \tag{9.93}$$

where R_E is defined by eqn. 5.22 with numerical values in Table 5.1. For long horns, as $l \to \infty$; $w \to 0$ and $R_E \to 1$ and $G = kb$, the exact gain of the resulting open-ended parallel-plate waveguide for TEM mode incidence, is obtained (see Section 9.38). Hence eqn. 9.93, which is also the Kirchhoff result, is an accurate approximation for long horns with small flare angles. However if singly diffracted fields are reflected from the horn interior in the axial direction, eqn. 9.87 rather than 9.84 is used in eqn. 9.36 and the gain is

$$G = (\pi/\phi_0) |\exp(-jkl \cos \phi_0) + 2v(l, \pi - \phi_0) + 2\exp(-jks_n)$$

$$\times v[l, \pi - (2m + 1)\phi_0] |^2 \tag{9.94}$$

Including double diffraction gives, for the axial gain of horns in which diffracted fields are not reflected in the forward direction,

$$G = (\pi/\phi_0) |\exp(-jkl \cos \phi_0) + 2v(l, \pi - \phi_0) + 2S |^2 \tag{9.95}$$

where S is defined by eqn. 9.88. If reflection in the forward direction can occur, $2 \exp(-jks_n)S^r$, with S^r given by eqn. 9.89, is added within the absolute value bars of eqn. 9.95)[†].

In Fig. 9.20 the gain per unit length of a two-dimensional E-plane sectoral horn with a total flare angle $2\phi_0 = 32°$ and an aperture width $b = 24$ cm is shown. The monotonic curve, based on eqn. 9.92 includes geometrical optics and singly diffracted fields from the aperture and is essentially the Kirchhoff result. The solid oscillating curve, based on eqn. 9.95 includes double diffraction in the horn. If triply diffracted fields are added, the dashed curve is obtained. It is evident these have little effect, except at the longer wavelengths. It is primarily the doubly diffracted fields which add the cyclic variation to the axial gain.

To use these results to account qualitatively for gain variations observed in pyramidal horns, as in Fig. 6.7, requires two assumptions. As a result of oblique incidence on the aperture edges the TE and TM-polarised interaction fields of a horn are coupled, but the TM-polarised interaction fields dominate in the axial direction and the observed gain variations are assumed to be entirely due to them. It is also reasonable to assume that only those TM fields from the waveguide which are incident on the central portions of the two E-plane aperture edges are diffracted in directions permitting reflection and double diffraction in the axial direction. This is consistent with Fermat's principle for diffracted rays. If all reflection and double diffraction of TM-polarised fields relevant to the axial gain is assumed to occur on an central portion of the E-plane edges equal in width to the waveguide feed ($|y| < a'/2$ in Fig. 6.2), then in the expressions for the axial gain these fields are reduced by a factor

$$\frac{\pi}{2a} \int_{-a'/2}^{a'/2} \cos\left(\frac{\pi y}{a}\right) dy = \sin\left(\frac{\pi a'}{2a}\right) \approx \frac{\pi a'}{2a}, \qquad a' \ll a$$

[†] The factor $\exp(-jks_n)$ is missing from the last term of eqn. 23 of Jull (1973b) and subsequently as a factor of S_2^r and S_3^r.

With these assumptions a revision of the pyramidal horn gain formula of eqn. 6.26 to include an approximation to double diffraction within the horn is

$$G = \frac{32ab}{\pi \lambda^2} R'_E R_H \qquad (9.96)$$

where R_H is defined as before by eqns. 5.29 and 6.22.

$$R'_E = \frac{1 + \cos \phi_0}{4w^2} \left| \exp(-jkl \cos \phi_0) + 2v(l, \pi - \phi_0) + \frac{\pi a'}{a} S \right|^2 \qquad (9.97)$$

in which $(1 + \cos \phi_0)/2$ appears from using eqn. 9.85 with $l = l_E$ for w, rather than eqn. 6.17. The final term in eqn. 9.97 accounts for double diffraction of the TM-polarised fields. If diffracted fields are reflected in the forward direction then

$$\frac{\pi a'}{a} \{v[l, \pi - (2m + 1)\phi_0] + \exp(-jks_n)S^r\}$$

is added within the absolute value bars of eqn. 9.97.

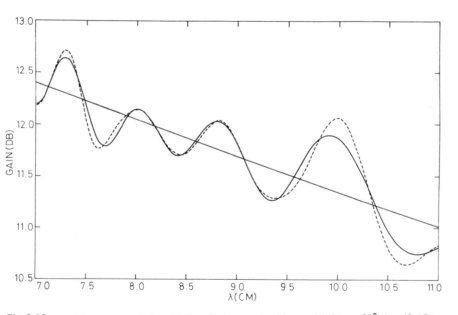

Fig. 9.20 *Axial gain per unit length of an E-plane sectoral horn with $2\phi_0 = 32°$, $l = 42\cdot15\,cm$ and $b = 24\cdot00\,cm$*
The non-interaction gain, eqn. 9.92, is nearly linear here. Eqn. 9.95 includes double diffraction and gives the solid oscillation. Dashed oscillation includes triple diffraction © 1973 IEEE (Jull, 1973b)

The oscillating solid curve of Fig. 6.7 is from eqn. 9.96 with R'_E defined by eqn. 9.97. It provides a qualitative explanation of the variations in the measured gain about values from Schelkunoff's equation (eqn. 6.26) represented by the solid monotonic curve, as well as an estimate of the error resulting from the use of (6.26) or (6.27). It fails to provide accurate quantitative values results because a two-dimensional solution is adequate here only for the non-interaction axial fields. Similar comparisons have been made for the other horns (Jull, 1973b).

9.5.5 Reflection from an E-plane sectoral horn

A long E-plane sectoral horn is one of the few aperture antennas for which the complex reflection coefficient and hence the impedance can be predicted. This requires the fields diffracted from the aperture and incident on the horn-waveguide junction. It is sufficient to consider only the dominant mode of these fields if the horn is sufficiently long and narrow that higher-order modes excited at the aperture are negligible at the throat.

If the incident field and the dominant mode of the field diffracted by the aperture edges of the two-dimensional E-plane sectoral horn of Fig. 9.21 are

$$H_z = \sqrt{\frac{\pi}{2}} \exp(-j\pi/4) \, [H_0^{(2)}(k\rho) + \alpha H_0^{(1)}(k\rho)]$$

$$E_\phi = -\frac{1}{j\omega\epsilon_0} \frac{\partial H_z}{\partial \rho} = Z_0\sqrt{\frac{\pi}{2}} \exp(-j\pi/4) \, [H_1^{(2)}(k\rho) + \alpha H_1^{(1)}(k\rho)]$$

$$(9.98)$$

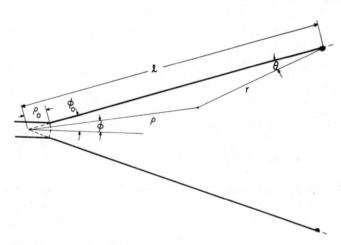

Fig. 9.21   *Coordinates in a two-dimensional sectoral horn*

and E_ρ, then α is required. For this it is convenient to determine the equivalent magnetic line source of the aperture edge diffracted fields.

With the first term above incident, the diffracted field of the upper edge within the horn of Fig. 9.21 is

$$H_z^d = \frac{\exp[-jk(l+r) + j3\pi/4]}{2\sqrt{kr} \, \sqrt{2\pi kl} \, \cos(\theta/2)}$$

$$(9.99)$$

or the asymptotic form of the second term of eqn. 9.67. Only rays diffracted at small angles θ reach the horn throat and, as eqn. 9.99 varies little for small θ, the diffracted fields there appear to originate from essentially isotropic line sources on horn walls of infinite extent. Hence for the aperture reflection coefficient it is convenient to let $\theta = 0$ in eqn. 9.99 and replace the edge by the line source

$$H_z^d = \frac{\omega \epsilon_0}{4} K H_0^{(2)}(kr) \tag{9.100}$$

where K is an equivalent magnetic current. Equating eqns. 9.99 and 9.100 and using the asymptotic form of the Hankel function eqn. 2.28, gives

$$K = \frac{j \exp(-jkl)}{\omega \epsilon_0 \sqrt{kl}} \tag{9.101}$$

for single diffraction. With the inclusion of double diffraction the magnetic current is KD where

$$D = 1 + \sqrt{\frac{2}{\pi}} \sum_{n=0}^{m} \frac{\exp[-jkd_n + j(3\pi/4)]}{\sqrt{kd_n}\,[1 + \sin(n+1)\phi_0]} \tag{9.102}$$

This follows from the asymptotic form of eqn. 9.79 with coordinates changed to those of Fig. 9.21 by omitting $\exp[j(kb/2)\sin\phi]$ and replacing ϕ by $\theta - \pi + \phi_0$. Then the approximation $\theta = 0$ is used as with singly diffracted fields.

The dominant mode fields in an infinite horn produced by eqn. 9.100 at $\rho = l$, $\phi = \phi_0$ can be written

$$H_z^d = \begin{cases} a_0 H_0^{(1)}(k\rho) + b_0 H_0^{(2)}(k\rho), & \rho_0 \leqslant \rho \leqslant l \\ c_0 H_0^{(2)}(k\rho), & \rho \geqslant l \end{cases} \tag{9.103}$$

where, from continuity of H_z^d at $\rho = l$,

$$c_0 = b_0 + a_0 \frac{H_0^{(1)}(kl)}{H_0^{(2)}(kl)} \tag{9.104}$$

The coefficient b_0 is determined from matching fields at the horn-waveguide junction and a_0 from the discontinuity in E_ϕ^d to give a line source at $\rho = l$, $\phi = \phi_0$. With eqn. 9.104 and the Wronskian formula

$$H_0^{(1)}(x)H_1^{(2)}(x) - H_1^{(1)}(x)H_0^{(2)}(x) = \frac{4j}{\pi x}$$

$$E_\phi^d(l_+) - E_\phi^d(l_-) = \frac{k}{j\omega\epsilon_0} \frac{a_0}{H_0^{(2)}(kl)} \left(\frac{4j}{\pi kl}\right) \tag{9.105}$$

$$= \frac{K}{2l} \delta(\phi - \phi_0) \tag{9.106}$$

where a factor of $1/2$ appears in eqn. 9.106 because the line source and its image coalesce on the conducting wall. Integrating eqns. 9.105 and 9.106 in ϕ from $-\phi_0$ to ϕ_0 gives

$$a_0 = \frac{\pi}{2\phi_0} \frac{\omega\epsilon_0}{8} K H_0^{(2)}(kl)$$

$$\simeq -\sqrt{\frac{\pi}{2}} \exp(-j\pi/4) \frac{\exp(-j2kl)}{8\phi_0 kl} \tag{9.107}$$

when eqn. 9.101 and the asymptotic form of $H_0^{(2)}(kl)$ are used. The coefficients of a full modal expansion are given elsewhere (Kinber, 1962; Jull, 1972).

As the line source at $\rho = l$, $\phi = -\phi_0$ yields the same result, the dominant mode coefficient for single diffraction by both edges is twice eqn. 9.107. Double diffraction is included by the factor D of eqn. 9.102. Hence in eqn. 9.98

$$\alpha = 2a_0 \simeq -\frac{\exp(-j2kl)}{4\phi_0 kl} D \tag{9.108}$$

for single and double diffraction at the aperture. Thus the aperture reflection coefficient of the horn is

$$\Gamma_s(\rho) = \frac{\alpha H_1^{(1)}(k\rho)}{H_1^{(2)}(k\rho)} = \frac{\exp\{-j2[kl - \phi(k\rho)]\}}{4\phi_0 kl} D \tag{9.109}$$

where $\phi(k\rho)$ is the phase angle of $H_1^{(1)}(k\rho)$.

If $k\rho \gg 1$, $\phi(k\rho) \simeq k\rho - 3\pi/4$, and the reflection coefficient at the aperture is

$$\Gamma_s(l) \simeq \frac{jD}{4\phi_0 kl} \tag{9.110}$$

This reduces to the open waveguide ($\phi_0 = 0$) solution (eqn. 9.17) by replacing ϕ_0 by $d_0/(2l)$ where d_0 is the plate spacing a of eqn. 9.17 and omitting all but the $n = 0$ term of the sum in eqn. 9.102, for double diffraction then occurs only across the aperture.

Calculation of this aperture reflection coefficient within the waveguide requires a further approximation as modal expressions within the horn and waveguide are not applicable in a common region (see Lewin, 1970). For small flare angles ϕ_0 the aperture reflection coefficient in the waveguide-horn junction is, approximately,

$$\Gamma_2 = \Gamma_s(\rho_0) \exp(-j\delta)$$

$$= -\frac{\exp\{-j2[kl - \phi(k\rho_0)] - j\delta\}}{4\phi_0 kl} D \tag{9.111}$$

where δ is a phase change due to the junction (Silver, 1949, Chap. 10).

$$\delta = 2 \tan^{-1} \left\{ \frac{A \cos[\psi(k\rho_0) - \phi(k\rho_0)]}{1 + A \sin[\psi(k\rho_0) - \phi(k\rho_0)]} \right\} \tag{9.112}$$

where

$$A = \frac{|H_0^{(1)}(k\rho_0)|}{|H_1^{(1)}(k\rho_0)|} \tag{9.113}$$

and $\psi(k\rho)$ is the phase angle of $H_0^{(1)}(k\rho)$.

This aperture reflection coefficient is superimposed on a junction reflection coefficient Γ_1. Matching dominant mode tangential fields at the horn waveguide junction gives, approximately,

$$\Gamma_1 = \frac{H_1^{(2)}(k\rho_0) - jH_0^{(2)}(k\rho_0)}{H_1^{(2)}(k\rho_0) + jH_0^{(2)}(k\rho_0)} \tag{9.114}$$

which applies for small ϕ_0.

If $|\Gamma_1\Gamma_s| \ll 1$ the total reflection coefficient in the waveguide is $\Gamma = \Gamma_1 + \Gamma_2$. Fig. 9.22 shows a plot of $|\Gamma|$ for a two-dimensional E-plane sectoral horn with $2\phi_0 = 33°$, $l = 34.7$ cm and $d_0 = 20$ cm. A comparison between the dashed oscillating curve, which includes double diffraction in Γ_2 and the solid curve, which does not, shows that double and higher-order diffraction is of minor importance, except for short horns for which the accuracy of Γ_2 is anyways limited.

Application of these results to the three-dimensional E-plane sectoral horn of Fig. 6.8a is straightforward if interaction between the two pairs of aperture edges can be neglected. For the closely spaced H-plane edges the TE_1 reflection coefficient of an open-ended parallel plate waveguide is required. Although the magnitude of the exact result of eqn. 9.23 is simple, its phase is not, and it is more convenient and consistant with other approximations to use the ray-optical result

$$R_{11} = \frac{j}{2k_1 a}\left(\frac{k}{k_1} - 1\right)\left\{1 + \sqrt{\frac{a}{\lambda}} \frac{\exp[-j(ka - 3\pi/4)]}{k_1 a}\right\} \tag{9.115}$$

with

$$k_1 = \frac{2\pi}{\lambda_g} = \sqrt{k^2 - (\pi/a)^2} \tag{9.116}$$

The first term in eqn. 9.115 represents single diffraction by the edges and is eqn. 9.22 with $n = N = 1$. The second term represents double diffraction. Together they approximate well the exact result for $0.6\lambda < a < \lambda$ (Yee *et al.*, 1968, Figs. 7a, b).

Although double diffraction by the long H-plane edges may be assumed essentially two-dimensional, it cannot for the short E-plane edges. However as the E-plane edges are well separated it is sufficient to include only the contribution of their singly diffracted fields to the reflection coefficient. Thus for the E-plane edges the reflection coefficient will be approximated by eqn. 9.111 with $D = 1$ and k replaced by k_1. Then if $|\Gamma_1\Gamma_2| \ll 1$, the reflection coefficient of a long E-plane sectoral horn is approximately

$$\Gamma = \Gamma_1 + \Gamma_2 \tag{9.117}$$

where the junction reflection coefficient is approximated by (e.g. Fradin, 1961, eqn. 5.130)

$$\Gamma_1 = \frac{H_1^{(2)}(k_1\rho_0) - jH_0^{(2)}(k_1\rho_0)}{H_1^{(2)}(k_1\rho_0) + jH_0^{(2)}(k_1\rho_0)} \tag{9.118}$$

The aperture contribution, obtained by adding the reflection coefficients of the edge pairs in phase, is

$$\Gamma_2 = -\exp\{-j2[k_1l - \phi(k_1\rho_0)] - j\delta\}\left\{\frac{1}{4\phi_0 k_1 l} + \frac{1}{2k_1 a}\left(\frac{k}{k_1} - 1\right)\right.$$

$$\left. \times \left[1 + 2\sqrt{\frac{a}{\lambda}}\frac{\exp[-j(ka - 3\pi/4)]}{k_1 a}\right]\right\} \tag{9.119}$$

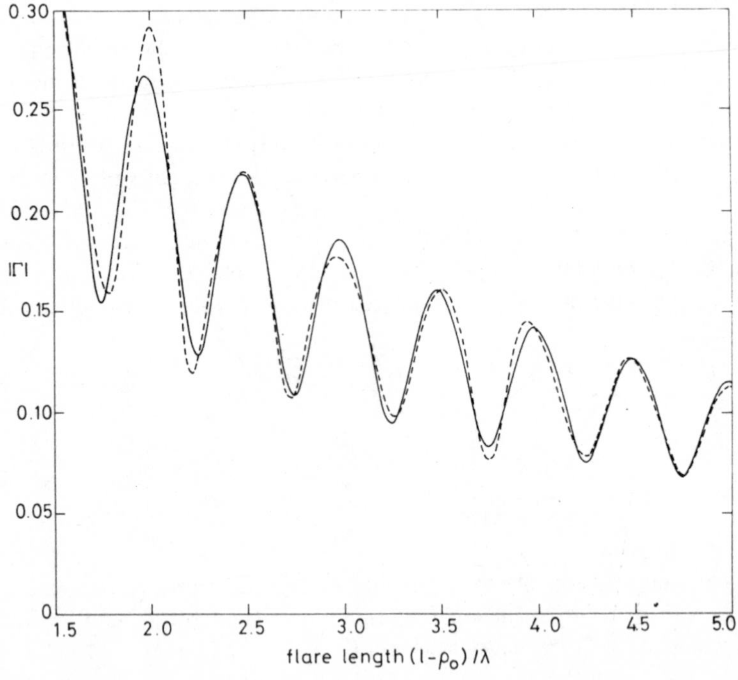

Fig. 9.22   *Calculated reflection coefficient of a two-dimensional E-plane sectoral horn with $2\phi_0 = 33.5°$, $l = 34.7\,cm$, $\rho_0 = 2.7\,cm$*
The dashed curve represents reflection from the junction and single and double diffraction at the aperture-eqns. 9.114 and 9.111. With a double diffraction not included ($D = 1$ in eqn. 9.111) the solid curve is obtained © 1972 IEEE (Jull, 1972)

in which δ is given by eqn. 9.112 with k replaced by k_1. The normalised admittance of the horn follows from $Y_N = (1 - \Gamma)/(1 + \Gamma)$.

Fig. 9.23 compares experimental and predicted reflection coefficients for an *E*-

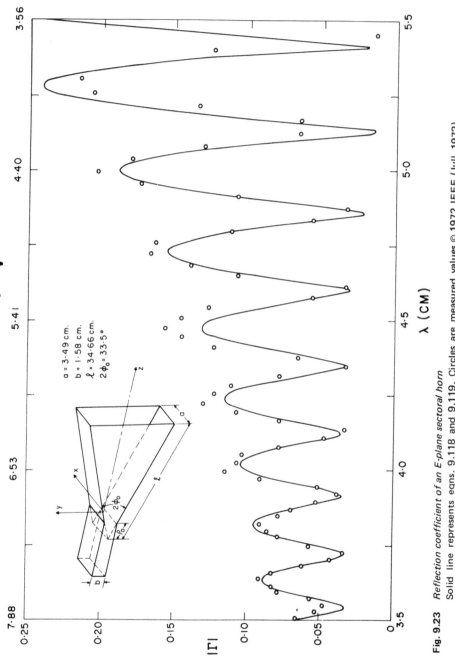

Fig. 9.23 *Reflection coefficient of an E-plane sectoral horn*
Solid line represents eqns. 9.118 and 9.119. Circles are measured values © 1972 IEEE (Jull, 1972)

plane sectoral horn. The component of the reflection coefficient which increases monotonically with wavelength is due to the horn-waveguide junction and is calculated from eqn. 9.118. The oscillating component, caused by aperture reflections, is from eqn. 9.119. Agreement is within the limits of experimental error here. Other comparisons with experiment (Jull, 1972) show that these expressions apply to E-plane sectoral horns when $(l - \rho_0)/\lambda_g$ is larger than about 3·5.

References

ABRAMOWITZ, M., and STEGUN, I. A. (1964): 'Handbook of mathematical functions', National Bureau of Standards, US Dept. of Commerce, reprinted 1965 (Dover)

AHLUWALIA, D. S., LEWIS, R. M., and BOERSMA, J. (1968): 'Uniform asymptotic theory of diffraction by a plane screen', *SIAM. J. Appl. Math.*, **16**, pp. 783–807

BABICH, V. M., and BULDYREV, V. S. (1972): 'Asymptotic methods of short wave diffraction', (in Russian) Moscow (Svyaz)

BAKER, B. B., and COPSON, E. T. (1950): 'The mathematical theory of Huygen's principle', 2nd edn. (Oxford)

BOERSMA, J. (1974): 'Ray-optical analysis of reflection in an open-ended parallel-plate waveguide. II-TE Case.' *Proc. IEEE*, **62**, pp. 1475–1481

BOERSMA, J. (1975a): 'Diffraction by two parallel half-planes', *Quart. J. Mech. Appl. Math.*, **38**, pp. 405–425

BOERSMA, J. (1975b): 'Ray-optical analysis of reflection in an open-ended parallel-plate waveguide. I-TM Case.' *SIAM. J. Appl. Math.*, **29**, pp. 164–195

BOERSMA, J., and LEE, S. W. (1977): 'High-frequency diffraction of a line source field by a half-plane: solutions by ray techniques', *IEEE Trans.*, **AP-25**, pp. 171–179

BOOKER, H. G. (1946): 'Slot aerials and their relationships to complementary wire aerials', *Proc. IEE*, **93IIIA**, pp. 620–626

BOOKER, H. G. and CLEMMOW, P. C. (1950): 'The concept of an angular spectrum of plane waves and its relation to that of polar diagram and aperture distribution', *Proc. IEE*, **97III**, pp. 11–17

BORGIOTTI, G. (1963): 'Radiation and reactive energy of aperture antennas', *IEEE Trans.*, **AP-11**, pp. 94–95

BORGIOTTI, G. (1967): 'On the reactive energy of an aperture', *IEEE Trans.*, **AP-15**, pp. 565–566

BORGIOTTI, G. V. (1978): 'Integral equation formulation for probe corrected far-field reconstruction from measurements on a cylinder', *IEEE Trans.*, **AP-26**, pp. 572–578

BORN, M., and WOLF, E. (Eds.) (1964): 'Principles of optics', 2nd edn., (Pergamon)

BOROVIKOV, V. A., and KINBER, B. Ye (1974): 'Some problems in the asymptotic theory of diffraction', *Proc. IEEE.*, **62**, pp. 1416–1437

BOROVIKOV, V. A., and KINBER, B. Ye (1978): 'Geometrical theory of diffraction', (in Russian) Moscow (Svyaz)

BOUWKAMP, C. J. (1946): 'A note on singularities at sharp edges in electromagnetic theory', *Physica*, **12**, pp. 467–474

BOUWKAMP, C. J. (1954): 'Diffraction theory', *Rept. Prog. Phys.*, **17**, pp. 35–100 (The Physical Society)

BOWMAN, J. J., SENIOR, T. B. A., and USLENGHI, P. L. E. (Eds.) (1969): 'Electromagnetic and acoustic scattering by simple shapes' (North-Holland)

BOWMAN, J. J. (1970): 'Comparison of ray theory with exact theory for scattering by open waveguides', *SIAM J. Appl. Math.*, **18**, pp. 818–829

BROWN, J. (1958a): 'A theoretical analysis of some errors in aerial measurements', *Proc. IEE*, **105C**, pp. 343–351

BROWN, J. (1958b): 'A generalised form of the aerial reciprocity theorem', *Proc. IEE*, **105C**, pp. 472–475

BROWN, J., and JULL, E. V. (1961): 'The prediction of aerial radiation patterns from near field measurements', *Proc. IEE*, **108C**, pp. 635–644

CHU, T. S., and SEMPLAK, R. A. (1965): 'Gain of electromagnetic horns', *Bell Syst. Tech. J.*, **44**, pp. 527–537

CIARKOWSKI, A. (1975): 'An application of uniform asymptotic theory of diffraction by a plane screen to the analysis of an open-ended parallel-plate waveguide', *Acta Phys. Polon.*, **A47**, pp. 621–631

CLEMMOW, P. C. (1950): 'A note on the diffraction of a cylindrical wave by a perfectly conducting half-plane', *Quart. J. Mech. Appl. Math.*, **3**, pp. 377–384

CLEMMOW, P. C. (1951): 'A method for the exact solution of a class of a two-dimensional diffraction problems', *Proc. Roy. Soc.*, **A205**, pp. 286–308

CLEMMOW, P. C. (1966): 'The plane wave spectrum representation of electromagnetic fields', (Pergamon)

COLLIN, R. E., and ROTHSCHILD, S. (1963): 'Reactive energy in aperture fields and aperture Q.', *Can J. Phys.*, **41**, pp. 1967–1979

COLLIN, R. E. (1967): 'Stored energy Q and frequency sensitivity of planar aperture antennas', *IEEE Trans.*, **AP-15**, pp. 567–568

COLLIN, R. E., and ZUCKER, F. J. (Eds.) (1969): 'Antenna theory Part I', (McGraw-Hill)

COPSON, E. T. (1946): 'On an integral equation arising in the theory of diffraction', *Quart, J. Math.*, **17**, pp. 19–34

CRAWFORD, A. B., HOGG, D. C., and HUNT, L. E. (1961): 'A horn reflector antenna for space communication', *Bell Syst. Tech. J.*, **40**, pp. 1095–1116

DRIESSEN, P. F. (1979): 'Ray theory of coupling between adjacent parallel-plate waveguides', *Radio Science*, **14**, pp. 969–978

FADDEYEVA, V. N., and TERENT'EV, N. M. (1961): 'Tables of the probability integral for complex argument', (Pergamon)

FELSEN, L. B., and YEE, H. Y. (1968): 'Ray-optical techniques for waveguide discontinuities', *IEEE Trans.*, **AP-16**, pp. 268–269

FELSEN, L. B., and MARCUVITZ, N. (1973): 'Radiation and scattering of waves', (Prentice-Hall)

FRADIN, A. Z. (1961): 'Microwave antennas', (Pergamon)

HANSEN, J. E. (1980): 'Spherical near-field testing of spacecraft antennas', *ESA J.*, **4**, pp. 89–102

HANSEN, R. C., and BAILLIN, L. L. (1959): 'A new method of near field analysis', *IRE Trans.*, **AP-7**, pp. S458–467

HANSEN, R. C. (1964): 'Aperture antennas', Chap. 1 *in* 'Microwave scanning antennas', vol. 1 (Academic)

HARRINGTON, R. F. (1961): 'Time-harmonic electromagnetic fields', (McGraw-Hill)

HEINS, A. E. (1948): 'The radiation and transmission properties of a pair of semi-infinite parallel-plates', *Quart, Appl. Math.*, **6**, part 1, pp. 157–166, part 2, pp. 215–220

ISHIHARA, T., FELSEN, L. B., and GREEN, A. (1978): 'High-frequency fields excited by a line source located on a perfectly conducting concave cylindrical surface', *IEEE Trans.*, **AP-26**, pp. 757–767

JAMES, G. L., and KERDEMELIDES, V. (1973): 'Reflector antenna analysis by equivalent edge currents', *IEEE Trans.*, **AP-21**, pp. 19–24

JAMES, G. L. (1974): 'Edge diffraction at a curved screen', *Electron. Lett.*, **10**, pp. 167–168

JAMES, G. L. (1976): 'Geometrical theory of diffraction for electromagnetic waves', IEE Electromagnetic Waves Series 1 (Peter Peregrinus)

JAHNKE, E., and EMDE, F. (1945): 'Tables of functions', (Dover)

JENSEN, F. (1975): 'On the probe compensation for near-field measurements on a sphere', *Arch. Electron. Übertragungstech*, **29**, pp. 305–308

JOHNSON, R. C., ECKER, H. A., and HOLLIS, J. S. (1973): 'Determination of far-field antenna patterns from near-field measurements', *Proc. IEEE*, **61**, pp. 1668–1694

JONES, D. S. (1979): 'Methods in electromagnetic wave propagation', Chap. 8 (Oxford)

JORDAN, E. C., and BALMAIN, K. G. (1968): 'Electromagnetic waves and radiating systems', 2nd edn. (Prentice-Hall)

JOY, E. B., LEACH, W. M., RODRIQUE, G. P., and PARIS, D. T. (1978): 'Applications of probe-compensated near-field measurements', *IEEE Trans.*, **AP-26**, pp. 379–389

JULL, E. V. (1962): 'An investigation of near-field radiation patterns measured with large antennas', *IRE Trans.*, **AP-10**, pp. 363–369

JULL, E. V. (1963): 'The estimation of aerial radiation patterns from limited near-field measurements', *Proc. IEE*, **110**, pp. 501–506

JULL, E. V., and DELOLI, E. P. (1964): 'An accurate absolute gain calibration of an antenna for radio astronomy', *IEEE Trans.*, **AP-12**, pp. 439–447

JULL, E. V. (1968a): 'Diffraction by a wide unidirectionally conducting strip', *Can. J. Phys.*, **46**, pp. 2107–2117

JULL, E. V. (1968b): 'On the behaviour of electromagnetic horns', *Proc. IEEE*, **56**, pp. 106–108

JULL, E. V. (1970): 'Finite range gain of sectoral and pyramidal horns', *Electron. Lett.*, **6**, pp. 680–681

JULL, E. V. (1972): 'Reflection from the aperture of a long E-plane sectoral horn', *IEEE Trans.*, **AP-20**, pp. 62–68

JULL, E. V. (1973a): 'Aperture fields and gain of open-ended parallel-plate waveguides', *IEEE Trans.*, **AP-21**, pp. 14–18

JULL, E. V. (1973b): 'Errors in the predicted gain of pyramidal horns', *IEEE Trans.*, **AP-21**, pp. 25–31

JULL, E. V., and ALLAN, L. E. (1974): 'Gain of an E-plane sectoral horn – a failure of the Kirchhoff theory and a new proposal', *IEEE Trans.*, **AP-22**, pp. 221–226

KARP, S. N., and RUSSEK, A. (1956): 'Diffraction by a wide slit', *J. Appl. Phys.*, **27**, pp. 886–894 [More numerical data is in Rep. EM-75, New York Univ. (1955)]

KARP, S. N., and KELLER, J. B. (1961): 'Multiple diffraction by an aperture in a hard screen', *Optica Acta*, **8**, pp. 61–72

KELLER, J. B. (1953): 'The geometrical optics theory of diffraction', McGill Symp. on Microwave Optics, AFCL Rep., TR-59-118 (II), 1959

KELLER, J. B. (1957): 'Diffraction by an aperture', *J. Appl. Phys.*, **28**, pp. 426–444

KELLER, J. B., LEWIS, R. M., and SECKLER, B. D. (1957): 'Diffraction by an aperture II', *J. Appl. Phys.*, **28**, pp. 570–579

KELLER, J. B. (1962): 'Geometrical theory of diffraction', *J. Opt. Soc. Am.*, **52**, pp. 116–130

KELLER, J. B., and HANSEN, E. B. (1965): 'Survey of the theory of diffraction of short waves by edges', *Acta Phys. Polon.*, **27**, pp. 217–234

KINBER, B. YE. (1961): 'Lateral radiation of parabolic antennas', *Radio Eng. and Elect. Phys.*, **6**, pp. 481–492

KINBER, B. YE. (1962): 'Diffraction at the open end of a sectoral horn', *Radio Eng. and Elect. Phys.*, **7**, pp. 1620–1632

KINBER, B. Ye., and POPICHENKO, V. A. (1972): 'Radiation from a sectoral horn', *Radio Eng. and Elect. Phys.*, **17**, pp. 1621–1627

KNOTT, E. F., and SENIOR, T. B. A. (1974): 'Comparison of three high-frequency diffraction techniques', *Proc. IEEE*, **62**, pp. 1468–1474

KOUYOUMJIAN, R. G., and PATHAK, P. H. (1974): 'A uniform geometrical theory of diffraction for an edge in a perfectly conducting surface', *Proc. IEEE*, **62**, pp. 1468–1461

KOUYOUMJIAN, R. G. (1975): 'The geometrical theory of diffraction and its application', Chap. 6 *in* 'Numerical and asymptotic techniques in electromagnetics', R. Mittra (Ed.) vol. 3, Topics in Appl. Phys (Springer)

KOUYOUMJIAN, R. G., and PATHAK, P. H. (1977): Author's reply to J. D. Cashman, 'Comments on a uniform geometrical theory of diffraction for an edge in a perfectly conducting surface:, *IEEE Trans.*, AP-25, pp. 449–451

KRAUS, J. D. (1950): 'Antennas', (McGraw-Hill)

LARSEN, Fl. H. (1977): 'Probe correction of spherical near-field measurements', *Electron. Lett.*, 13, pp. 393–395

LEACH, W. M., and PARIS, D. T. (1973): 'Probe-compensated near field measurements on a cylinder', *IEEE Trans.*, AP-21, pp. 435–445.

LEACH, W. M., LARSEN, F. H., and BORGIOTTI, G. V. (1979): 'Comments on "Integral equation formulation for probe corrected far-field reconstruction from measurements on a cylinder" ' *IEEE Trans.*, AP-27, pp. 895–898

LEE, S. W. (1969: 'On edge diffraction rays of an open-ended waveguide', *Proc. IEEE*, 57, pp. 1445–1446

LEE, S. W. (1970): 'Ray theory of diffraction by open-ended waveguides. I Field in waveguides', *J. Math. Phys.*, 11, pp. 2830–2850

LEE, S. W., and BOERSMA, J. (1975): 'Ray-optical analysis of fields on shadow boundaries of two parallel plates', *J. Math. Phys.*, 16, pp. 1746–1764

LEE, S. W., and DESCHAMPS, G. (1976): 'A uniform asymptotic theory of electromagnetic diffraction by a curved wedge', *IEEE Trans.*, AP-24, pp. 25–34

LEE, S. W. (1978): 'Uniform asymptotic theory of electromagnetic edge diffraction: a review' *in* 'Electromagnetic scattering', Uslenghi, P. L. E. (Ed.) (Academic) pp. 67–119

LEWIN, L. (1969): 'Wedge diffraction functions and their use in quasioptics', *Proc. IEE*, 116, pp. 71–76

LEWIN, L. (1970): 'On the inadequacy of discrete mode-matching techniques in some waveguide discontinuity problems', *IEEE Trans.*, MTT-18, pp. 364–369

LEWIN, L. (1972): 'Main reflector-rim diffraction in back direction', *Proc. IEE*, 119, pp. 1100–1102

LEWIN, L. (1976): 'Some uses of Huygen's obliquity factor', *Radio Science*, 11, pp. 445–448

LEWIS, R. M. and BOERSMA, J. (1969): 'Uniform asymptotic theory of edge diffraction', *J. Math. Phys.*, 10, pp. 2291–2305

LOVE, A. W. (1978) (Ed.), 'Reflector antennas', (IEEE Press)

MARTIN, W. W. (1967): 'Computation of antenna radiation patterns from near-field measurements', *IEEE Trans.*, AP-15, pp. 316–318

MEIXNER, J. (1954): 'The behaviour of Electromagnetic fields at edges' Rep. EM-72, New York Univ., reprinted (1972) *IEEE Trans.*, AP-20, pp. 442–446

MENTZER, C. A., PETERS, L., and RUDDUCK, R. C. (1975): 'Slope diffraction and its application to horns', *IEEE Trans.*, AP-23, pp. 153–159

MEYER, C. F. (1934): 'The diffraction of light, X-rays and material particles', (U. Chicago)

MILLAR, R. F. (1956): 'An approximate theory of the diffraction of an electromagnetic wave by an aperture in a plane screen', *Proc. IEE*, 103C, pp. 177–185

MILLAR, R. F. (1957): 'The diffraction of an electromagnetic wave by a circular aperture', *Proc. IEE*, 104C, pp. 87–95

MILLAR, R. F. (1958): 'Diffraction by a wide slit and complementary strip. I and II', *Proc. Camb. Phil. Soc.*, 54, pp. 479–511

MILNE, K. (1952): 'The effects of phase errors on simple aperture distributions'. Proceedings of a conference on centimetric antennas for marine navigational radar. London, June 1950, HMSO, pp. 86–122

MITTRA, R., RAHMAT-SAMII, Y., and KO, W. L. (1976): 'Spectral theory of diffraction', *Appl. Phys.*, 10, pp. 1–13

NARASIMHAN, M. S., and VENTKATESWARA, RAO, V. (1973): 'A correction to the available radiation formula for E-plane sectoral horns', *IEEE Trans.*, AP-21, pp. 878–879.

NOBLE, B. (1958): 'Methods based on the Wiener-Hopf technique for the solution of partial differential equations', (Pergamon)

OHBA, Y. (1961): 'Geometrical method of diffraction', *Trans. Prof. Group on Field Theory*, Japan, No. 34

OHBA, Y. (1962): 'The radiation pattern of a parabolic reflector antenna in the lateral and backward directions', *Proc of the Fujihara Mem. Fac. of Eng.*, Keio, Univ., Tokyo, **15**, No. 59, pp. 1–7

OHBA, Y. (1963): 'On the radiation pattern of a corner reflector finite in width', *IEEE Trans.*, **AP-11**, pp. 127–132

PARIS, D. T., LEACH, W. M., and JOY, E. B. (1978): 'Basic theory of probe-compensated near-field measurements', *IEEE Trans.*, **AP-26**, pp. 379–389

PAULI, W. (1938): 'On asymptotic series for functions in the theory of diffraction of light', *Phys. Rev.*, **54**, pp. 924–931

PENZIAS, A. A., and WILSON, R. W. (1965): 'A measurement of excess antenna temperature at 4080 Mc/s.', *Astrophys. J.*, **142**, pp. 419–421

POLK, C. (1956): 'Optical Fresnel-zone gain of a rectangular aperture', *IRE Trans.*, **AP-4**, pp. 65–69

RAHMAT-SAMII, Y., and MITTRA, R. (1977): 'A spectral domain interpretation of high-frequency diffraction phenomena', *IEEE Trans.*, **AP-25**, pp. 676–687

RAMSAY, J. F. (1946-47): 'Fourier transforms in aerial theory', *Marconi Rev.*, **83**, pp. 139–145; **84**, pp. 17–22; **85**, pp. 41–58; **86**, pp. 81–90; **87**, pp. 157–165

RHODES, D. R. (1964): 'On a fundamental principle in the theory of planar antennas', *IEEE Trans.*, **AP-12**, pp. 1013–1021

RHODES, D. R. (1966): 'On the stored energy of planar apertures', *IEEE Trans.*, **AP-14**, pp. 676–683

RHODES, D. R. (1974): 'Synthesis of planar antenna sources' (Oxford)

RUBINOWICZ, A. (1924): 'Zur Kirchhoffschen beugungstheorie', *Ann. Phys.*, **73**, pp. 339–364

RUBINOWICZ, A. (1957): 'Thomas Young and the theory of diffraction', *Nature*, **180**, pp. 160–162

RUDDUCK, R. C. (1965): 'Application of wedge diffraction to antenna theory', Electroscience Lab. Rep. 7691-13, Ohio State Univ.

RUDDUCK, R. C., and TSAI, L. L. (1968): 'Aperture reflection coefficients of TEM and TE_{01} mode parallel-plate waveguides', *IEEE Trans.*, **AP-16**, pp. 83–89

RUDDUCK, R. C., and WU, D. C. F. (1969): 'Slope diffraction analysis of TEM parallel-plate guide patterns', *IEEE Trans.*, **AP-17**, pp. 797–799

RUSCH, W. V. T., and SØRENSEN, O. (1975): 'The geometrical theory of diffraction for axially symmetric reflectors', *IEEE Trans.*, **AP-23**, pp. 414–419

RUSSO, P. M., RUDDUCK, R. C., and PETERS, L. (1965): 'A method for computing E-plane pattern of horn antennas', *IEEE Trans.*, **AP-13**, pp. 219–224

RYAN, C. E., and PETERS, L. (1969): 'Evaluation of edge-diffracted fields including equivalent currents for caustic regions', *IEEE Trans.*, **AP-17**, pp. 292–299

SAFAAI-JAZI, A., and JULL, E. V. (1977): 'A short horn with high E-plane directivity', *IEEE Trans.*, **AP-25**, pp. 854–859

SCHELKUNOFF, S. A. (1943): 'Electromagnetic waves', (Van Nostrand)

SCHWARZSCHILD, K. (1902): 'Die beugung und polarization des lichts durch einem spalt. I', *Math. Ann.*, **55**, pp. 177–247

SILLER, C. A. (1975): 'Evaluation of the radiation integral in terms of end-point contributions', *IEEE Trans.*, **AP-23**, pp. 743–745

SILVER, S. (Ed.) (1949): 'Microwave antenna theory and design', MIT Rad. Lab. Ser. Vol. 12, (McGraw-Hill)

SKAVLEM, S. (1951): 'On the diffraction of scalar plane waves by a slit of infinite length', *Arch. Math. Nat.*, **B51**, pp. 61–80

SLAYTON, W. T. (1954): 'Design and calibration of microwave antenna gain standards', US NRL Rep. 4433

SMITH, J. M. (1963): 'A note on diffraction theory and polarization', *Proc. IEE.*, **110**, pp. 88–90

SOEJIMA, T. (1963): 'Fresnel gain of aperture aerials', *Proc. IEE*, **110**, pp. 1021–1027

SOMMERFELD, A. (1896): 'Mathematische theorie der diffraction', *Math. Ann.*, **47**, pp. 317–374 [See also 'Optics' (1965), pp. 245–265 (Academic)]

SPENCER, R. C., and AUSTIN, P. M. (1946): 'Tables and methods of calculation for line sources', MIT Rad. Lab. Rep. 762–2

STRATTON, J. A. (1941): 'Electromagnetic theory', (McGraw-Hill)

THOUREL, L. (1960): 'The antenna', (Wiley)

TSAI, L. L., WILTON, D. R., HARRISON, M. G., and WRIGHT, E. H. (1972): 'A comparison of geometrical theory of diffraction and integral equation formulation for analysis of reflector antennas', *IEEE Trans.*, **AP-20**, pp. 705–712

UFIMTSEV, P. Ya (1962): 'Method of edge waves in the physical theory of diffraction' (in Russian), Iz. Sov. Radio, pp. 1–243. Translation (1971) by USAF Foreign Tech. Div. Wright-Patterson AFB

UFIMTSEV, P. YA (1975): 'Comments on "Comparison of three high-frequency diffraction techniques"', *Proc. IEEE*, **63**, pp. 1734–1737

YANKOUGHNETT, A. L., and WONG, J. Y. (1971): 'H-plane radiation patterns of a dipole antenna mounted above a finite rectangular ground screen', Rep. ERB-852, Div. Electr. Eng., Nat. Res. Council, Ottawa

WACKER, P. F. (1975): 'Non-planar near-field measurements: spherical scanning', Rep. NBS IR 75-809, Nat. Bur. Stds., Boulder. Co.

WAIT, J. R. (1953): 'Radiation from a line source adjacent to a conducting half-plane', *J. Appl. Phys.*, **24**, pp. 1529–1530

WEINSTEIN, L. A., (1948): 'Rigorous solution of the problems of an open-ended parallel-plate waveguide' (in Russian), *Izv. Akad. Nauk. SSSR*, Ser. Fiz. 12, pp. 144–165

WEINSTEIN, L. A. (1966): 'The theory of diffraction and the factorization method', Chap. 1 (Golem)

WHITTAKER, E. T., and WATSON, G. N. (1920): 'A course of modern analysis', 3rd edn., (Cambridge)

WOOD, P. J. (1980): 'Reflector antenna analysis and design', IEE Electromagnetic Waves Series 7 (Peter Peregrinus)

WOODWARD, P. M. (1946): 'A method of calculating the field over a plane aperture required to produce a given polar diagram', *Proc. IEE*, **93 III**, pp. 1554–1558

WOODWARD, P. M., and LAWSON, J. D. (1948): 'The theoretical precision with which an arbitrary radiation pattern may be obtained from a source of finite size', *Proc. IEE*, **95III**, pp. 363–370

YEE, H. Y., FELSEN, L. B., and KELLER, J. B. (1968): 'Ray theory of reflection from the open end of a waveguide', *SIAM J. Appl. Math.*, **16**, pp. 268–300

YEE, H. Y., and FELSEN, L. B. (1968): 'Ray optical techniques for waveguide discontinuities', Rep. PIBEP-68-005, Polytech. Inst. Brooklyn, NY

YU, J. S., RUDDUCK, R. C., and PETERS. L. (1966): 'Comprehensive analysis for E-plane of horn antennas by edge diffraction theory', *IEEE Trans.*, **AP-14**, pp. 138–149

YU, J. S., and RUDDUCK, R. C. (1967): 'Higher-order diffraction concepts applied to a conducting strip', *IEEE Trans.*, **AP-15**, pp. 662–668

YU, J. S., and RUDDUCK, R. C. (1969): 'H-plane pattern of a pyramidal horn', *IEEE Trans.*, **AP-17**, pp. 651–652

Appendixes

A.1 Fourier integrals and transforms

If a function $f(\xi)$ and its derivative are piecewise continuous in the range $-l/2, l/2$ it may be expanded in a Fourier series.

$$f(\xi) = a_0 + \sum_{n=1}^{\infty} \left[a_n \cos\left(\frac{2\pi n\xi}{l}\right) + b_n \sin\left(\frac{2\pi n\xi}{l}\right) \right]$$

$$= \sum_{n=-\infty}^{\infty} c_n \exp\left(j\frac{2\pi n\xi}{l}\right) \tag{A.1}$$

where $c_0 = a_0$ and $c_n = \frac{1}{2}(a_n \mp jb_n), n \gtrless 0$.

The coefficients c_n are obtained by multiplying both sides by $\exp(-j2\pi m\xi/l)$ and integrating term by term in ξ over the complete range. Then

$$c_m = \frac{1}{l} \int_{-l/2}^{l/2} f(\xi) \exp\left[-j(2\pi m\xi/l)\right] d\xi \tag{A.2}$$

The limit of the Fourier series as the period tends to infinity is the Fourier integral. If, in eqn. A.1, $2\pi n/l$ and c_n are replaced by η and $g(\eta)(2\pi/l)$, respectively, the difference between successive values of η written $d\eta$ and the limits of eqn. A.2 extended to $(-\infty, +\infty)$, eqns. A.1 and A.2 become

$$f(\xi) = \int_{-\infty}^{\infty} g(\eta) \exp(j\xi\eta) d\eta \tag{A.3}$$

and

$$g(\eta) = \frac{1}{2\pi} \int_{-\infty}^{\infty} f(\xi) \exp(-j\eta\xi) d\xi \tag{A.4}$$

These are known as a Fourier transform pair, $g(\eta)$ being the spectrum function of $f(\xi)$. Other forms are also used. For example, if in eqn. A.3 ξ is replaced by $kS = 2\pi S/\lambda$, the transforms become

$$f(kS) = \int_{-\infty}^{\infty} g(\eta) \exp(jkS\eta)\,d\eta$$

$$g(\eta) = \frac{1}{\lambda}\int_{-\infty}^{\infty} f(kS)\exp(-jkS\eta)\,dS \tag{A.5}$$

A symmetrical form is obtained by putting $\xi = 2\pi u$ in (A.3), giving

$$f(2\pi u) = \int_{-\infty}^{\infty} g(\eta)\exp(j\pi u\eta)\,d\eta$$

$$g(\eta) = \int_{-\infty}^{\infty} f(2\pi u)\exp(-j2\pi u\eta)\,du \tag{A.6}$$

The Dirac delta function $\delta(\xi) = 0$, $\xi \neq 0$, must be defined in terms of integration. Usually this is

$$\int_{-\infty}^{\infty} \delta(\xi)\,d\xi = 1 \tag{A.7}$$

and $\delta(\xi)$ has the property that, for any $f(\xi)$,

$$\int_{-\infty}^{\infty} f(\xi)\,\delta(\xi)\,d\xi = f(0) \tag{A.8}$$

Hence, replacing $f(\xi)$ by $\delta(\xi)$ in eqn. A.4 gives its Fourier transform $g(\eta) = 1/2\pi$. Using this in eqn. A.3 gives an alternative definition

$$\delta(\xi) = \frac{1}{2\pi}\int_{-\infty}^{\infty} \exp(j\xi\eta)\,d\eta \tag{A.9}$$

A.2 Stationary phase evaluation of integrals

A.2.1 Single integrals
Consider

$$I = \int_{a}^{b} f(\alpha)\exp[-jkrg(\alpha)]\,d\alpha \tag{A.10}$$

where $f(\alpha)$ and $g(\alpha)$ are real and $kr \gg 1$. The integrand oscillates rapidly along the integration path except where

$$\frac{\partial}{\partial \alpha} g(\alpha) = g'(\alpha) = 0$$

with the solution $\alpha = \alpha_0$ called the 'stationary point'. This phase interference causes cancellation and

$$I \approx \int_{\alpha_0 - \delta}^{\alpha + \delta} f(\alpha) \exp[-jkrg(\alpha)] d\alpha, \qquad 0 < \delta \ll 1$$

Here

$$f(\alpha) \approx f(\alpha_0) + 0(\delta)$$

$$g(\alpha) \approx g(\alpha_0) + \frac{g''(\alpha_0)}{2}(\alpha - \alpha_0)^2 + 0(\delta^3)$$

since $g'(\alpha_0) = 0$. Hence

$$I \approx f(\alpha_0) \exp[-jkrg(\alpha_0)] \int_{\alpha_0 - \delta}^{\alpha_0 + \delta} \exp[-j(kr/2) g''(\alpha_0)(\alpha - \alpha_0)^2] d\alpha$$

Let

$$\xi = \sqrt{\frac{kr}{2} g''(\alpha_0)} (\alpha - \alpha_0)$$

Then

$$I \approx \sqrt{\frac{2}{krg''(\alpha_0)}} f(\alpha_0) \exp[-jkrg(\alpha_0)] \int_{-\xi_0}^{\xi_0} \exp(-j\xi^2) d\xi$$

where

$$\xi_0 = \sqrt{\frac{kr}{2} g''(\alpha_0)} \delta$$

The limits can be extended to $\pm \infty$ with negligible error and

$$\int_{-\infty}^{\infty} \exp(-j\xi^2) d\xi = \sqrt{\pi} \exp(-j\pi/4)$$

used giving

$$I \simeq \sqrt{\frac{2\pi}{kr|g''(\alpha_0)|}} f(\alpha_0) \exp[-jkrg(\alpha_0) \mp j\pi/4] \qquad \text{(A.11)}$$

the $\mp$ sign being chosen accordingly as $g''(\alpha_0) \gtrless 0$; i.e. the root

$$\sqrt{g''(\alpha_0)} = -j\sqrt{|g''(\alpha_0)|} \text{ is used for } g(\alpha_0) < 0.$$

A.2.2. Double integrals

The integral to be evaluated for $kr \gg 1$ is

$$I = \int \int f(\alpha,\beta) \exp[-jkrg(\alpha,\beta)] \, d\alpha d\beta \tag{A.12}$$

The stationary point is $\alpha = \alpha_0$, $\beta = \beta_0$, solutions of

$$\frac{\partial g}{\partial \alpha} = \frac{\partial g}{\partial \beta} = 0$$

Expanding $g(\alpha,\beta)$ in a Taylor's series in two variables about the stationary point gives

$$g(\alpha,\beta) = g(\alpha_0,\beta_0) + \tfrac{1}{2}a(\alpha-\alpha_0)^2 + \tfrac{1}{2}b(\beta-\beta_0)^2$$

$$+ c(\alpha-\alpha_0)(\beta-\beta_0) + \ldots$$

where

$$a = \frac{\partial^2 g}{\partial \alpha^2}\bigg|_{\substack{\alpha=\alpha_0\\\beta=\beta_0}}, \quad b = \frac{\partial^2 g}{\partial \beta^2}\bigg|_{\substack{\alpha=\alpha_0\\\beta=\beta_0}}, \quad c = \frac{\partial^2 \beta}{\partial \alpha \partial \beta}\bigg|_{\substack{\alpha=\alpha_0\\\beta=\beta_0}}$$

With $\xi = \alpha - \alpha_0$, $\eta = \beta - \beta_0$,

$$g(\alpha,\beta) = g(\alpha_0,\beta_0) + \tfrac{1}{2}a\xi^2 + \tfrac{1}{2}b\eta^2 + c\xi\eta$$

Then

$$I \simeq f(\alpha_0,\beta_0) \exp[-jkrg(\alpha_0,\beta_0)] \int \int \exp[-j(kr/2)(a\xi^2 + b\eta^2 + 2\xi\eta c)] \, d\xi d\eta$$

$$= \frac{-2\pi j\sigma}{kr\sqrt{|ab-c^2|}} f(\alpha_0,\beta_0) \exp[-jkrg(\alpha_0,\beta_0)] \tag{A.13}$$

where

$$\sigma = 1 \qquad ab > c^2, \quad a > 0$$
$$= -1 \qquad ab > c^2, \quad a < 0$$
$$= j \qquad ab < c^2$$

See also Silver (1949, pp. 119–122) and Born and Wolf (1964, pp. 752–754).

A.3 Integral and asymptotic forms of the Hankel functions

With a time dependence $\exp(j\omega t)$, the zero-order Hankel function of the second kind

$$H_0^{(2)}(kr) = J_0(kr) - jN_0(kr) \tag{A.14}$$

is a complex representation of an outgoing cylindrical wave. It can also be defined in integral form as the field due to a uniform line distribution of isotropic sources. In the coordinates of Fig. A.1

$$H_0^{(2)}(kr) = \frac{j}{\pi} \int_{-\infty}^{\infty} \frac{\exp(-jk\rho)}{\rho} dz \qquad (A.15)$$

is a linear superposition of elementary spherical waves. The substitution $z = r \sinh \tau$ in eqn. A.15 gives

$$H_0^{(2)}(kr) = \frac{j}{\pi} \int_{-\infty}^{\infty} \exp(-jkr \cosh \tau) d\tau \qquad (A.16)$$

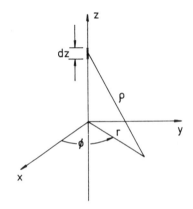

Fig. A.1   *Coordinates for radiation from a line source*

and, on letting $\tau = -j\zeta$,

$$H_0^{(2)}(kr) = \frac{1}{\pi} \int_{-j\infty}^{j\infty} \exp(-jkr \cos \zeta) d\zeta \qquad (A.17)$$

The integration contour C_0 is along the imaginary axis of the complex $\zeta = \xi + j\eta$ plane, as in Fig. A.2.

In eqn. A.17, since

$$\text{Re}(jkr \cos \zeta) = kr \sin \xi \sinh \eta > 0 \begin{cases} 0 < \xi < \pi, & \eta > 0 \\ -\pi < \xi < 0, & \eta < 0 \end{cases}$$

the integrand decays exponentially in the shaded strips and the integration contour C_0 can be deformed into a contour such as C_1, without altering the value of the integral (by Cauchy's theorem). Thus

$$H_0^{(2)}(kr) = \frac{1}{\pi} \int_{C_1} \exp(-jkr\cos\zeta)d\zeta \tag{A.18}$$

is a convenient integral representation of the Hankel function.

For large kr, the asymptotic form of the Hankel function may be obtained by a stationary phase evaluation of the integral. According to Section A.2.1., the stationary point is obtained from $g'(\zeta) = -\sin\zeta = 0$ and is at $\zeta = 0$ on C_1. Also, $g''(\zeta = 0) = -1$ and, from eqn. A.11,

$$H_0^{(2)}(kr) \simeq \sqrt{\frac{2}{\pi kr}} \exp[-jkr + j(\pi/4)] \tag{A.19}$$

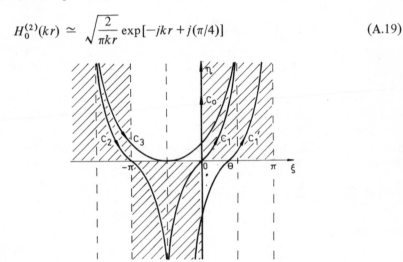

Fig. A.2   *Integration contours in the complex ζ-plane*

an exponential form of cylindrical wave. Hence, according to eqns. A.14 and A.19,

$$\frac{J_0(kr)}{N_0(kr)} \simeq \sqrt{\frac{2}{\pi kr}} \frac{\cos}{\sin}\left(kr - \frac{\pi}{4}\right) \tag{A.20}$$

and the relationship between Bessel functions and the Hankel function for cylindrical waves is equivalent to that between the cosine and sine functions and the exponential function for plane waves.

Letting $\zeta = \alpha - \theta$, where θ is real, in eqn. A.18 simply shifts the contour C_1 to C_1' in Fig. A.2 and

$$H_0^{(2)}(kr) = \frac{1}{\pi} \int_{C_1'} \exp[-jkr\cos(\alpha - \theta)]d\alpha \tag{A.21}$$

has the asymptotic form of eqn. A.19. Hence the angular spectrum of outgoing plane waves of equal amplitude $1/\pi$ is a Hankel function of zero order.

Because $H_0^{(2)}(kr)$ is a cylindrical wave or linear superposition of elementary plane or spherical waves it must be a solution of the two-dimensional wave equation which depends on only the distance r from the z-axis, i.e. Bessel's equation of order zero:

$$\frac{\partial^2 u}{\partial r^2} + \frac{1}{r}\frac{\partial u}{\partial r} + k^2 u = 0 \tag{A.22}$$

which is a linear differential equation of the second order. It must have two linearly independent solutions, of which $H_0^{(2)}(kr)$ is one. A second is its complex conjugate

$$H_0^{(1)}(kr) = J_0(kr) + jN_0(kr) \tag{A.23}$$

$$= \int_{C_2} \exp(-jkr\cos\zeta)d\zeta \tag{A.24}$$

$$\simeq \sqrt{\frac{2}{\pi kr}}\exp[jkr - j(\pi/4)] \tag{A.25}$$

which represents an incoming cylindrical wave for $\exp(j\omega t)$ time dependence. The integration contour C_2 is chosen to pass through the stationary point at $\zeta = -\pi$ in evaluating eqn. A.24 for $kr \gg 1$.

Cylindrical fields which vary in ϕ as well as r must contain linear combinations of solutions of the Bessel equation of order n:

$$\frac{\partial^2 u}{\partial r^2} + \frac{1}{r}\frac{\partial u}{\partial r} + [k^2 - (n/r)^2]u = 0 \tag{A.26}$$

or, for example,

$$H_n^{(1)}(kr) = J_n(kr) + jN_n(kr) \tag{A.27}$$

$$H_n^{(2)}(kr) = J_n(kr) - jN_n(kr) \tag{A.28}$$

Integral forms of these Hankel functions may be obtained from the zero order functions. For example,

$$H_1^{(2)}(kr) = -\frac{\partial}{\partial(kr)}H_0^{(2)}(kr)$$

$$= \frac{j}{\pi}\int_{C_1}\exp(-jkr\cos\zeta)\cos\zeta\,d\zeta$$

$$= \frac{j}{\pi}\int_{C_2}\exp(-jkr\cos\zeta - j\zeta)d\zeta$$

In general,

$$H_n^{(2)}(kr) = \frac{\exp[jn(\pi/2)]}{\pi}\int_{C_1}\exp(-jkr\cos\zeta - jn\zeta)d\zeta \tag{A.29}$$

$$\simeq \sqrt{\frac{2}{\pi k r}}\, \exp[-jkr + j(2n + 1)\pi/4] \tag{A.30}$$

An integral form of $H_n^{(1)}(kr)$ is eqn. A.29 with the integration contour along C_2 of Fig. A.2. Its asymptotic form is the complex conjugate of eqn. A.30.

A general two-dimensional field $\partial/\partial z = 0$ may consist of a sum of mode solutions to

$$\frac{\partial^2 u}{\partial r^2} + \frac{1}{r}\frac{\partial u}{\partial r} + \frac{1}{r}\frac{\partial^2 u}{\partial \theta^2} + k^2 u = 0 \tag{A.31}$$

i.e.

$$u = \sum_{-\infty}^{\infty} a_n H_n^{(2)}(kr)\exp(jn\theta) \tag{A.32}$$

which is complex form of Fourier series expansion of a two-dimensional field in cylindrical coordinates.

A useful integral form of the Bessel function follows from eqns. A.27 and A.28:

$$J_n(kr) = \tfrac{1}{2}[H_n^{(1)}(kr) + H_n^{(2)}(kr)]$$

$$= \frac{\exp[jn(\pi/2)]}{2\pi}\int_{C_2 + C_1} \exp(-jkr\cos\zeta - jn\zeta)\,d\zeta \tag{A.33}$$

in which $C_2 + C_1$ may be replaced by the contour C_3 of Fig. A.2, or any contour containing a segment of length 2π along the real axis. Its asymptotic form is the real part of eqn. A.30.

A.4 Solution of the dual integral equations of half-plane diffraction

Here the dual integral equations 7.21 and 7.22 are solved by the contour integration techniques used by Clemmow (1951, 1966) but with additional mathematical justification. The exponential factor in the integrands can be written as $\exp(-jk_x'x + k_x''x)$, where $k_x' = \mathrm{Re}(k_x)$ and $k_x'' = \mathrm{Im}(k_x)$. Then, since $x < 0$ in eqn. 7.22, the integration contour may be closed by an infinite arc above the integration path of Fig. 7.3 without further contribution if, according to Jordan's lemma (e.g. Whittaker and Watson, 1920, p. 115),

$$U(k_x) = \sqrt{k^2 - k_x^2}\, F(k_x) \tag{A.34}$$

is analytic there and if

$$U(k_x) \to 0 \text{ as } |k_x| \to \infty,\ k_x'' > 0 \tag{A.35}$$

That $U(k_x)$ is analytic follows from its definition in terms of the surface current density $J_z(x)$. From eqns. 7.2 and 7.16 this current density is

$$J_z(x) = H^s_{x,+} - H^s_{x,-}$$

$$= -\frac{2}{\omega\mu_0} \int_{-\infty}^{\infty} k_y F(k_x) \exp(-jk_x x) dk_x \tag{A.36}$$

and consequently

$$U(k_x) = k_y F(k_x) = -\frac{\omega\mu_0}{4\pi} \int_{-\infty}^{\infty} J_z(x) \exp(jk_x x) dx \tag{A.37}$$

Since $J_z(x)$ is continuous for all $x \geq 0$, the entire integrand of eqn. A.37 is continuous and analytic for all k_x and $x \geq 0$. Also, it converges as $|k_x| \to \infty$, $k''_x > 0$ since $x \geq 0$ in eqn. A.37. Thus $U(k_x)$ is analytic above the path of integration (Whittaker and Watson, 1920, p. 92).

To show eqn. A.35, we observe that in eqn. A.37, $J_z(x) = 0(x^{-1/2})$ as $x \to 0_+$, since this current component is proportional to the magnetic field H^s_x which has this behaviour (see eqn. 7.4). Hence as $|k_x| \to \infty$, $k''_x > 0$, in eqn. A.37 the dominant contribution to the integral occurs near its lower integration limit. Since

$$\int_0^\infty x^\beta \exp(jk_x x) dx = \frac{\Gamma(\beta+1)}{(-jk_x)^{\beta+1}}, \qquad \beta > -1 \tag{A.38}$$

$U(k_x) = 0(k_x^{-1/2}) \to 0$ as $|k_x| \to \infty$, $k''_x \to 0$, verifying eqn. A.35.

The integration contour of eqn. 7.21, where $x > 0$, can be closed by an infinite semicircle below the path of integration of Fig. 7.3 if $F(k_x)$ has a simple pole at $k_x = -k_0$ and is otherwise free of singularities below the integration path. From eqns. 7.15 and 7.19,

$$F(k_x) = \frac{1}{2\pi} \int_{-\infty}^{0} E_z(x,0) \exp(jk_x x) dx - \delta(k_x + k_0) \tag{A.39}$$

where the delta function $\delta(k_x + k_0)$ (see eqn. A.9) represents the pole. The integrand in eqn. A.39 is continuous and analytic for all k_x and converges as $|k_x| \to \infty$, $k''_x < 0$ since $x \leq 0$. As $E_z(x,0) = 0(x^{1/2})$ as $x \to 0_-$, $F(k_x)$ satisfies the criteria for closure of the contour below the path of integration and eqn. 7.21 satisfies Cauchy's theorem if

$$F(k_x) = \frac{1}{2\pi j} \frac{L(k_x)}{L(-k_0)} \frac{1}{k_x + k_0} \tag{A.40}$$

when the path of integration is indented above the pole at $k_x = -k_0$ as in Fig. 7.3.

Here $L(k_x)$ is a function analytic and of no more than algebraic growth at infinity below the path of integration.

If eqn. A.40 is rewritten as

$$\frac{U(k_x)}{\sqrt{k-k_x}}(k_x + k_0) = \frac{1}{2\pi j}\frac{L(k_x)}{L(-k_0)}\sqrt{k+k_x} \qquad (A.41)$$

then the left-hand side is free of singularities above the integration path of Fig. 7.3 and the right-hand side is free of singularities below it. Also $U(k_x) = 0(k_x^{-1/2})$ and hence $U(k_x)(k_x + k_0)(k-k_x)^{-1/2} = 0(k_x^0)$ as $|k_x| \to \infty$ above the real axis. The right-hand side behaves similarly below it. Hence both sides must equal a function analytic and bounded over the entire complex k_x-plane, which by Liouville's theorem (e.g. Whittaker and Watson, 1920, p. 105) is a constant. This constant is obtained by inserting $k_x = -k_0$ in the right-hand side of eqn. A. 41. Then

$$U(k_x) = \frac{1}{2\pi j}\frac{\sqrt{k-k_x}\sqrt{k-k_0}}{k_x + k_0}$$

and

$$F(k_x) = \frac{U(k_x)}{\sqrt{k^2 - k_x^2}} = \frac{1}{2\pi j}\sqrt{\frac{k-k_0}{k+k_x}}\frac{1}{(k_x + k_0)} \qquad (A.42)$$

It is worth noting that the problem may also be formulated as an integral equation of the Wiener-Hopf type by using eqn. A.37 in eqns. 7.21. On reversing the order of the integration, letting $k_x = k\cos\zeta$ and recognising the integral form of the Hankel function eqn. A.18 one obtains

$$\frac{kZ_0}{4}\int_0^\infty J_z(x') H_0^{(2)}(k|x-x'|)dx' = \exp(jk_0x), \quad x > 0 \qquad (A.43)$$

a Wiener-Hopf integral equation which can be solved by standard techniques. (e.g. Noble, 1958).

A.5 Reduction of the half-plane diffraction solution to Fresnel integrals

With $k_x = k\cos\alpha$ and $k_0 = k\cos\theta_0$, eqn. 7.24 becomes, for $y \gtrless 0$,

$$E_z^s = \frac{1}{\pi j}\int_C \frac{\sin(\alpha/2)\sin(\theta_0/2)}{\cos\alpha + \cos\theta_0}\exp[-jkr\cos(\theta \mp \alpha)]\,d\alpha \qquad (A.44)$$

$$= \frac{1}{4\pi j}\int_C \left[\sec\left(\frac{\alpha + \theta_0}{2}\right) - \sec\left(\frac{\alpha - \theta_0}{2}\right)\right]\exp[-jkr\cos(\theta \mp \alpha)]\,d\alpha$$

$$(A.45)$$

which is the form of solution used by Clemmow (1951, 1966). Following his procedure, the path of integration C of Fig. A.3 is deformed into the steepest descents path $S(\theta)$ through the saddle point at $\alpha = \theta$, making allowance for any poles captured in the process. For $y \geqslant 0$, the upper signs in eqn. A.45 apply and there is a pole at $\alpha = \pi - \theta p$, which from Fig. A.3 is crossed when $\theta < \pi - \theta_0$, yielding $-\exp[(jkr \cos(\theta + \theta_0)]$ by Cauchy's theorem. This is the reflected wave of geometrical optics.

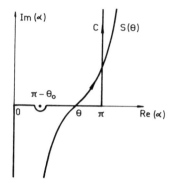

Fig. A.3 *Integration contours in the complex α-plane*

It is apparent from eqn. A.45 that it is sufficient to evaluate integrals of the form

$$I = \int_{S(\theta)} \sec\left(\frac{\alpha + \theta_0}{2}\right) \exp[-jkr \cos(\theta - \alpha)]\, d\alpha \tag{A.46}$$

as the result for the second term in eqn. A.45 follows by changing the sign of θ_0. The new variable $\alpha' = \alpha - \theta$ in eqn. A.46 yields

$$I = \int_{S(0)} \sec\left(\frac{\alpha' + \theta_0 + \theta}{2}\right) \exp(-jkr \cos\alpha')\, d\alpha'$$

$$= \frac{1}{2} \int_{S(0)} \left[\sec\left(\frac{\alpha' + \theta_0 + \theta}{2}\right) + \sec\left(\frac{\alpha' - \theta_0 - \theta}{2}\right)\right] \exp(-jkr \cos\alpha')\, d\alpha'$$

$$= 2\cos\left(\frac{\theta + \theta_0}{2}\right) \int_{S(0)} \frac{\cos(\alpha'/2) \exp(-jkr \cos\alpha')}{\cos\alpha' + \cos(\theta + \theta_0)}\, d\alpha' \tag{A.47}$$

Let $t = \sqrt{2} \exp(-j\,\pi/4)\sin(\alpha'/2)$. Then along the path of integration t goes through real values from $-\infty$ to ∞. Now $\cos\alpha' = 1 - jt^2$, so for $kr \gg 1$ the exponential term in eqn. A.47 causes the integrand to decrease rapidly about $t = 0$. With this substitution

$$I = -2\exp[-j(kr + \pi/4)]b \int_{-\infty}^{\infty} \frac{\exp(-krt^2)}{t^2 + jb^2} \, dt \qquad \text{(A.48)}$$

where $b = \sqrt{2}\cos[(\theta + \theta_0)/2]$. Eqn. A.48 is a form of Fresnel integral which can be rewritten as (Clemmow, 1966, p. 52)

$$I = \mp 4\sqrt{\pi}\exp(-j\pi/4)\exp[jkr\cos(\theta + \theta_0)]\, F\!\left[\pm\sqrt{2kr}\cos\!\left(\frac{\theta + \theta_0}{2}\right)\right]$$
$$\text{(A.49)}$$

for $\cos[(\theta + \theta_0)/2] \gtrless 0$, i.e. $\theta \lessgtr \pi - \theta_0$. In eqn. A.49, the complex Fresnel integral is defined by eqn. 7.26.

The result for the second term in eqn. A.45 has no sign ambiguity as both α and θ_0 are in the range θ, π. Combining the results, eqn. A.45 becomes

$$E_z^s = \frac{\exp(j\pi/4)}{\sqrt{\pi}}\left\{ \pm\exp[jkr\cos(\theta + \theta_0)]\, F\!\left[\pm\sqrt{2kr}\cos\!\left(\frac{\theta + \theta_0}{2}\right)\right]\right.$$
$$\left. - \exp[jkr\cos(\theta - \theta_0)]\, F\!\left[\sqrt{2kr}\cos\!\left(\frac{\theta - \theta_0}{2}\right)\right]\right\} \qquad \text{(A.50)}$$

depending on whether $\theta \lessgtr \pi - \theta_0$, plus the pole contribution $-\exp[jkr\cos(\theta + \theta_0)]$ for $\theta < \pi - \theta_0$. Eqn. 7.27 can be used to combine both pole contribution and incident field eqn. 7.14 with eqn. A.50 to yield eqn. 7.25 for the total field.

When eqn. A.45 is evaluated for $y < 0$, $(\pi < \alpha < 2\pi)$, the pole at $\alpha = \pi + \theta_0$ contributes when $\theta > \pi + \theta_0$ and yields minus the incident wave. The complete field is given by eqn. 7.25, which is valid for θ and θ_0 in the entire range $0, 2\pi$.

A.6 Transmission cross-section of a slit

The transmission cross-section of an aperture is defined as the ratio of the power transmitted through the aperture to the power intercepted by the aperture. It can be obtained in a simple way from the spectrum function or the radiation pattern of the aperture.

To illustrate, consider diffraction of a TE-polarised plane wave incident from $y > 0$ on a slit $|x| < a/2$ in conducting screen in $y = 0$. The incident field is given by eqn. 7.14 and the total fields in $y < 0$ by expressions of the form of eqns. 7.15–7.16. The power transmitted through the slit per unit length in the z-direction is obtained by integrating the normal component of the complex Poynting vector over the aperture plane. This is

$$\frac{1}{2}\operatorname{Re}\int_{-\infty}^{\infty}\bar{E}\times\bar{H}^{*}\cdot\hat{n}\,dx = -\frac{1}{2}\operatorname{Re}\int_{-a/2}^{a/2}E_{z}H_{x}^{*}dx \tag{A.51}$$

since $E_z = 0$, $|x| > a/2$. Also, since $H_x = H_x^i$ in $y = 0$, $|x| < a/2$, using eqn. 7.14, eqn. A.51 becomes

$$\frac{1}{2}Y_0\sin\theta_0\operatorname{Re}\int_{-a/2}^{a/2}E_z(x,0)\exp(-jk_0x)\,dx \tag{A.52}$$

where $k_0 = k\cos\theta_0$.

If the *total* fields in $y \leqslant 0$ are represented in the form of eqns. 7.15–7.16, the spectrum function $F(k_x)$ is related to the total electric field by

$$F(k_x) = \frac{1}{2\pi}\int_{-a/2}^{a/2}E_z(x,0)\exp(jk_xx)\,dx \tag{A.53}$$

again since $E_z = 0$ in $y = 0$ for $|x| > a/2$. Hence

$$F(-k_0) = \frac{1}{2\pi}\int_{-a/2}^{a/2}E_z(x,0)\exp(-jk_0x)\,dx \tag{A.54}$$

which from eqn. A.52 is proportional to the power transmitted through the slit. Dividing eqn. A.52 by $\frac{1}{2}Y_0$, or the power density of the incident field in the direction of incidence, and using eqn. A.54, the transmission cross-section of the slit is

$$\sigma = 2\pi\sin\theta_0\operatorname{Re}\{F(-k_0)\}$$
$$= 2\pi\sin\theta_0\operatorname{Re}\{F(k\cos(\pi+\theta_0))\} \tag{A.55}$$

Hence the transmission cross-section is proportional to the value of the spectrum function in the forward direction, i.e. the direction of propagation of the incident wave $\theta = \pi + \theta_0$. To express σ in terms of the pattern function $f(\phi)$ of Section 8.3 requires a stationary phase evaluation of

$$E_z = \int_{-\infty}^{\infty}F(k_x)\exp[-j(k_xx+k_yy)]\,dk_x$$

for $kr \gg 1$. Referring to Section 2.5, this is

$$E_z = -2\pi\sin\theta F(k\cos\theta)\frac{\exp[-j(kr-\pi/4)]}{\sqrt{\lambda r}}$$
$$= f(\phi)\frac{\exp[-j(kr+\pi/4)]}{\sqrt{\lambda r}} \tag{A.56}$$

where

$$f(\phi) = -2\pi j \sin \theta F(k \cos \theta) \qquad (A.57)$$

is the pattern function and $\phi = 3\pi/2 - \theta$ as in Fig. 8.2. Then eqn. A.57 in eqn. A.55 gives

$$\sigma = \mathrm{Im}\{f(\theta_0 - \pi/2)\} \qquad (A.58)$$

This result is of general validity for any array of slits or for a TM-polarised incident field. In the latter case, the radiation pattern $f(\phi)$ is that associated with the component of the *total* field parallel to the incident field $\bar{\mathbf{E}}^i$.

The scattering cross-section of an obstacle is defined as the ratio of the total power in the *scattered* field to the power in the incident field. For a strip in $y = 0$, $|x| < a/2$ with a plane wave incident in the direction θ_0, the scattering cross-section is, from Babinet's principle,

$$\sigma = -2\,\mathrm{Im}\{f(\theta_0 - \pi/2)\} \qquad (A.59)$$

where the additional factor 2 enters because scattered fields on both sides of the strip are included. In eqn. A.59, the pattern $f(\phi)$ is that associated with the component of *scattered* field $\bar{\mathbf{E}}^s$ parallel to the incident field $\bar{\mathbf{E}}^i$.

Author index

Index